Desirable Futu

CW01486347

Desirable Future?

Consumer Electronics in Tomorrow's World

Jack Challoner

Other Wiley Editorial Offices

John Wiley & Sons Inc., 111 River Street, Hoboken, NJ 07030, USA

Jossey-Bass, 989 Market Street, San Francisco, CA 94103-1741, USA

Wiley-VCH Verlag GmbH, Boschstr. 12, D-69469 Weinheim, Germany

John Wiley & Sons Australia Ltd, 42 McDougall Street, Milton, Queensland 4064, Australia

John Wiley & Sons (Asia) Pte Ltd, 2 Clementi Loop #02-01, Jin Xing Distripark, Singapore 129809

John Wiley & Sons Canada Ltd, 6045 Freemont Blvd, Mississauga, ONT, L5R 4J3, Canada

Wiley also publishes its books in a variety of electronic formats. Some content that appears in print may not be available in electronic books.

British Library Cataloguing in Publication Data
A catalogue record for this book is available from the British Library

ISBN 978-0-470-986608

Typeset in 9.5 on 14 pt SM DIN by SNP Best-set Typesetter Ltd., Hong Kong
Printed and bound by Printer Trento in Italy.

Contents

Introduction

If you are anything like me, then consumer electronics fill you with desire: you don't need those gadgets, you want them. You want to play with them, to marvel at the ingenuity of their invention, and enjoy and appreciate their design. It will feel good 'unboxing them' and plugging them in. You want to own them so that you can interact with them, bring them into your life. That desire is the anticipation of pleasure.

While writing this book, my desire for the latest high-tech gadgets has been heightened, because I have been more in touch than ever with the latest developments in consumer electronics. As a result – and for research purposes of course – I have upgraded my mobile phone; I have bought a pocket media player with a large touch screen; I have bought a high-definition flat-screen television and hung it on the wall; I have bought a single-lens reflex digital camera; and I have bought a new laptop computer.

All of my shiny new devices have satisfied my desire to a certain extent, but I know that in six months time – or probably sooner – there will be something else to rekindle it. In fact, there are already things I would love to have. Ultra-light laptops that you can connect directly to the Internet through mobile broadband via a built-in antenna, even when there is no WiFi hotspot to be found? Brilliant: I want one. But I don't need one – not really. In the same way, I didn't really have to upgrade my mobile phone: the old one worked just fine and had several features I didn't even use. But it was satisfying to get my hands on a newer model, with many of the latest features. Likewise, I didn't really need a large flat-screen television: our living room is small, and our existing television was adequate and produced a good picture. But I wanted to experience the clarity of high-definition television for myself and buy into the slick, futuristic, hang-it-on-the-wall mentality. I didn't really need an SLR digital camera: the smaller, cheaper digital camera I had before more than satisfied my snap-happy nature. And that camera on its own had already produced more lovely photos than I could organise in my busy life. And the feat of organising my digital lifestyle is made harder because I have changed my computer, where I store all those photos along with music and videos, at least twice in the past five years. But for all that, I wanted to take pictures with better resolution and use a camera with more manual features.

Okay, so I'm a tech junkie. However, despite what my list of recent purchases suggests, I am not a rich tech junkie. Ordinarily, I don't get my fix very often, and like most people, I have to be very careful how I spend my money. I still don't own a PlayStation 3 or an X-Box Live, nor a Wii games console, much as I desire to have all three. I decided not to go for an Apple iPhone – but only because the tariff was too high for me to justify. I haven't got a Blu-ray or HD-DVD player, surround sound or wireless speakers – though I would love to treat our new television (and myself) to all of them. The list of unfulfilled desires is, I am sure, quite long. But please don't feel sorry for me. I feel guilt for the things I do buy: guilt about the electrical power these devices use; guilt about the intensive use of materials to make them and the waste of materials when I upgrade; and guilt about the fact that a large proportion of the world's population cannot afford to have these things. My guilt would be so much worse if I could afford to feed my habit more often – by buying the next generation of console or computer chip as soon as it came out. If I had more money, I would be torn between the desire to upgrade and the guilt of spending all that money on stuff I don't really need.

Taking such an interest in the latest consumer electronics gadgets, I am always eager to know what is coming next. Today's gadgets are amazing – what will tomorrow's ones be like? How will they work? How might we use them? I am curious to see just what is possible. If you are like me, then I hope this book will answer these questions for you. In case you are not like me – if instead of desire, consumer electronics gadgets fill you with dread or simply apathy – then this book should still be of interest to you. Firstly, if you are a bit of a technophobe, I hope that understanding how things work will make them less daunting, or at least a bit more interesting. Secondly, our relationship with and our reliance on digital tools and playthings seem set to grow – so the future of consumer electronics affects us all. Any book about the future is, of course, speculative. You might think that fact would be particularly true for a book about consumer electronics, because the technology moves ahead so quickly. In fact, as we will see, the next ten years at least are fairly well charted, and even many of the possible developments beyond that are already well thought out.

In Chapter 1, to prepare you for your journey into the future, I explain the principles behind the amazing technologies involved in the digital revolution. However marvellous are our current technologies, they have not delivered us into utopia. The reality of consumer electronics at the moment is populated by expensive devices that sometimes do not communicate easily with each other; by batteries that run out of charge far too quickly and often just at the wrong time; by homes littered with tangles of spaghetti-like wires; by gadgets that are too complicated for many people to use effectively, and which are rarely exploited to their full potential; and by services that are not always available if you are anywhere outside a major city. There are gadgets that 'crash' at inopportune moments, forcing you to remove a battery or find a paper clip with which to push the 'reboot' button. And then there is the cost of staying up to date – to benefit from what the latest gadgets have to offer and to ensure that your devices are not becoming obsolete. Many people find rapid progress in technology very exciting, but it can also be very frustrating.

Rather than putting people off consumer electronics, however, this imperfect and fragmented state of affairs is one of the reasons why so many people are looking forward to tomorrow's world. In other words, part of the appeal of the next big thing is the recognition of the limitations of today's technologies and services. There is a real desire for faster, smaller, more 'connected' and more powerful devices. And that desire is one of the main drivers of technological change. It helps fuel the cycle of research and development, marketing and consumer uptake that characterises the modern consumer electronics industry. Chapter 2 takes a look at that lucrative industry, which cleverly manages to satisfy our desire while also keeping it alive.

Chapter 3 examines the most important and sustained technological trends behind the digital revolution and goes on to predict a major shift in our relationship with what is currently the centre of our digital lives, the personal computer. There are limitations inherent in the current approach to design and manufacturing, which threaten to slow down and halt progress in consumer electronics in the next ten years. Chapter 4 considers how this might lead to new ways of interacting with our gadgets, and how more intelligent gadgets should be able to communicate more naturally with us and anticipate our needs. Chapter 5 explains the emerging technologies that will enable the trends in consumer electronics to defy these limitations and continue for decades to come. It also examines other technological advances that will help to shape the landscape of consumer electronics and determine what kinds of gadgets will be available in the next ten to twenty years. With technological progress more or less assured, we can foresee a future in which integrated, connected and affordable consumer electronics will be the norm – so much so that they will probably blend into the backgrounds of our lives.

In *The Future of Mobility*, a film produced for the Japanese mobile services provider NTT DoCoMo, we hear that in the near future 'not only people, but everything, will be connected in a world network of wireless connections that will heighten convenience and enrich all of society.' This kind of phrase is reminiscent of the naïve techno-optimism widespread in the 1950s – in which technological progress brings nothing but benign happiness. Such optimism is usually devoid of the gritty realism of everyday life – and as a result, feels unconvincing. Furthermore, the kinds of futures portrayed in these optimistic visions are nearly always based on the lifestyles of executives or otherwise wealthy people. Someone is nearly always booking the next flight to Tokyo by talking to his or her watch strap, or an intelligent electronic personal assistant is sending three-dimensional digital architectural models to its boss's ultra-thin laptop in time for the morning board meeting, or else a team of personal robots is hosting a dinner party in a plush mansion. Such visions rarely show these technologies as they would be used by 'normal' people – although that is not to say that such people will not benefit from many of the technologies portrayed in these visions.

By their nature, optimistic portrayals of the future tend to ignore the darker sides of technological progress: the environmental cost of consumer electronics, the digital divide between the haves and the have-nots, the potential negative consequences to our health and our social relations, and the increasing threats to our privacy and security. Technology can do good or bad things – it is important to remember that both are possible. In Chapter 6, we explore some of the less talked about aspects of consumer electronics, and we consider some potential delays or diversions on the road map to the kind of future many people assume is just around the corner. These include the possibility that consumer capitalism itself may falter under pressure from spiralling energy costs, over-sophistication, terrorism or simple consumer apathy.

It is unlikely, even after global catastrophes or a severe dressing down of consumer capitalism – perhaps by a growing anti-globalisation movement – that personal, portable, connected electronic gadgets will disappear altogether. In Chapter 7, we consider what our personal gadgets might be like, and what they may do for us, in the more distant future. Such is the nature of humankind that tomorrow's technology will more than likely surpass anything in even the most optimistic visions – eventually.

Prelude: Tomorrow and Today

It is 2030. The science fiction stories and predictions of the middle of the 20th century were spot on, just as you assumed they would be, and you are living in a peaceful yet exciting utopia, in which intelligent, sophisticated, but easy-to-use electronic devices integrate seamlessly into your life at every turn. They are in your pocket, they are embedded in your clothes, and they surround you in your home and at work. They serve you intelligently, unobtrusively and often invisibly. You speak to them, they respond. There are immersive 3D video games with lifelike graphics – as good as being there; wall-sized video displays; flexible, handheld, paper-thin devices that allow you to hold video chats with your friends, family or colleagues anywhere in the world. Images appear in your eyes like magic, and sounds fill your ears. Your car navigates and drives itself, responding to your spoken requests of where you want to go and constantly monitoring traffic and weather conditions.

Electronic technology is vital in every area of your life in 2030 – education and government, work and play, travel and medicine. The emphasis is on choice and convenience; technology has removed all those chores from your daily and weekly routines – unless you want to tackle them. So, for example, when you receive bills and bank statements, they are of no immediate concern to you: the computer takes care of your finances, noticing any trends of which you need to be aware, and communicating them to you at opportune moments. At any time, you can ask to be presented with the information, and take as active an interest in your finances as you wish. But generally, you let the computer get on with it. Likewise, a wealth of monitors both inside and outside your body constantly measure your vital signs, including such things as blood glucose level and hormone concentrations, and communicate them to your house computer, which acts as your personal doctor, alerting you to anything of note, and referring you to a human doctor when necessary. This is a world in which digital entertainment, information convenience and unhindered communication are intimately woven into the fabric of everyday life. Anything seems possible, and everything is effortless.

It is morning; the voice of your house computer gently wakes you – close to the time you wanted to wake, but the computer has taken into account your sleep cycle, to ensure that you will be as alert as possible. The computer adjusts the amount of light coming in through the window. You open one eye and demand that the windows show a view of the ocean from a pebbly beach. The computer obliges, instantly displaying an exquisite image, complete with a golden sunrise and accompanied by the sounds of waves and a gentle breeze. It is as if you really live in a beachfront house, just for today. You don't inform the computer that you are having a shower; but it detects your intention to do so and turns on the water, which attains the perfect temperature just as you step into the cubicle.

As you dry and dress, the computer informs you of any messages and news that you are likely to find important. It filters the urgent and desirable from the annoying and irrelevant. You take a three-dimensional video call from your work colleague – the image hovers in the air next to the wall, and follows you into the bathroom and back again as you get ready for the day. Afterwards, you ask what the weather is likely to do today, and the computer answers accordingly and automatically displays a local weather map on the large, wall-sized video screen that is as thin as wallpaper. You announce what you would like for breakfast, and as you wander into the kitchen the robot there is preparing it for you. At the same time, the house computer is checking the inventory of what is left in the fridge and the kitchen cupboards, in preparation for ordering the week's shopping. As you sit eating your breakfast, your eldest child looks up and grunts a morning greeting. He is busy chatting to three of his friends whose faces appear on the living room wall. You take in more news and messages on the other wall, interacting with the images and text using your voice and hand gestures.

When it is time to leave the house, you pick up your thin but robust mobile computer-communicator. There is no keyboard: you control the gadget with touch and voice. You can even mouth the words – detectors in your clothes can work out what you are saying by sensing which muscles in your throat are moving. And sensors in the gadget take note of subtle gestures and eye movements to help discern your intentions. All of this makes the whole experience completely intuitive and very convenient – but you are used to that: this is 2030, after all. The device remains in constant communication with your home computer without your intervention. It is hard to lose this gadget, because it constantly communicates its position to your home computer – or indeed any other device that is connected to the Internet. And if you do lose it or forget it, you can get another one anyway; these devices are cheap, universal and light – and biodegradable. Unlike the mobile phones, media players or laptop computers from way back when, they carry hardly any information inside. Instead, these computer-communicators are simply 'front-end' interfaces to your vast store of your information on your house computer.

And so, as you leave the house for work, the house locks automatically, and the house computer's security features take care of everything while you are away. No keys are needed; your voice and your fingerprints will suffice. En route to work or play, whatever music or video, news or communication you require is available to you in fantastic quality.

I know, I know: next you'll be jumping into a hovering taxi wearing weird silver boots and a faux leather cat suit. And you'll be going to a party in a pod-shaped house with white walls and brushed steel-and-shiny-black butler robots that shuffle around bringing drinks. This kind of tame and tired portrayal of the future feels oversimplified, sterile – and overplayed. But there is an important point: this kind of slick, convenience-filled future is quietly assumed by many people to be the destination towards which our technologically advanced world is heading. And it is seemingly well within our grasp – a mere extrapolation from today – because it relies on technologies that are either possible now or that are currently in development. There is certainly nothing in what I have portrayed that is impossible. It is difficult to tell whether such a world would be a wonderful place to live or a terrible place filled with in-your-face unavoidable advertising, covert surveillance of all citizens, corruption and greed. There is no way from this kind of prissy vignette to tell whether everyone would have this kind of lifestyle and luxury, or whether the less well-off would live a watered-down version.

People who presently use the latest consumer gadgets are living in a prequel to that future vision. Already, thanks to powerful digital electronics, you can use portable devices to speak to friends, family and business colleagues anywhere, at any time. You can play exciting video games – even competing online with other gamers far away, in real time. You can watch a huge range of on-demand television programming in exquisite high-definition quality on stylish, thin, flat screens. You can capture precious moments of your life – as recorded sounds, still images and video – and share them with friends and family quickly and easily. And you can use your devices to help you shop and book travel or concert tickets from anywhere. You can even buy a watch that monitors the state of your sleep cycle by measuring your heart rate and body temperature, and wakes you at the optimum moment – the lightest phase of sleep within a desired window of time. No hovering taxis yet, though.

However many individual elements of the high-tech future may already be available today, we are certainly not there yet. Consider a scenario from today's world. It highlights how much a part of our lives consumer electronics have become, but it also hints at the disconnectedness of our gadgets, and some of their annoying foibles.

Today's World

It is 2009. You wake in the morning to the foghorn-like sound of the alarm on your mobile. You 'snooze' it repeatedly, but after the third time of asking, you drag yourself out of bed and into the day. Once you are dressed, you watch the breakfast news, although you come in at the end of a report that seemed interesting. You could go online and find out what it was about, but it would take a while to boot up your PC and find the relevant information. You make sure you have your mobile phone and your keys, and you head off to the train. Your fellow passengers are busily thumbing text messages or emails, or talking into their mobile phones. Some are watching videos or listening to podcasts on their personal media players or perhaps even reading a novel on their e-book reader. Many of them are reading newspapers and magazines . . . made of paper. You content yourself with listening to some music and reading through some work-related documents.

At work, you log in to your computer, and you switch between email, web browser and word processing programs. You check the quarantined emails that have been labelled as spam, and find that two of them were actually important messages from work contacts. Saving things to and retrieving them from the server seems to take ages, and once in a while, your computer crashes. Later in the day, you have outside meetings – on the way, you get a lift with your colleague, who finds the venue using a portable GPS unit – through which she also makes voice-activated, hands-free phone calls. She has to drive herself, though – there is no autonomous robot driver; and at your destination, she must remember to unplug the GPS unit and hide it in her bag. At the meeting, a laptop is used with a projector, and at the end you wirelessly swap electronic business cards with others at the meeting – although not all the mobiles are compatible with each other. At the end of the day, you ride home on the train; you flip open your mobile phone and, despite the small screen, you watch a television programme. The quality and the choice of programmes are not quite as you would like, but you pay a lot for this service, so you feel you should make the most of it – despite the small screen.

When you get home, you boot up your PC, rid your own email inbox of spam, and reply to an email from your relatives abroad, attaching recent photos and a short piece of video of your family. The Internet seems annoyingly slow, as usual. Your eldest child is logged on to one of his favourite social networking sites, exchanging instant messages with his friends. You sit and watch your LCD television, perhaps playing back the programme you set to record on your personal video recorder. Afterwards, you enjoy a video game on your games console, competing against other gamers far away via your home's Internet connection. At the end of the evening, you remind yourself to plug in your portable devices to charge their batteries ready for the next day, knowing that the alarm on your mobile phone will wake you in the morning.

Many of these experiences will be familiar to most of us – whether or not we travel to work on the train and have office-based jobs. The range of products available, as well as rapidly falling prices, means that most of us possess several of these gadgets. The rise of consumer electronics has radically changed the face of entertainment, information and communication. A 2005 report by Intel notes that 'almost unnoticed over the past few years, consumers have undergone a profound change in the way they view the role that technology plays in their lives.' Just fifteen years ago, most households would have had only a television, a telephone, a video cassette recorder and a home stereo system, which would probably have included a CD player. Back then, many people saw mobile phones as symbols of wealth or importance, and ownership was rare. Go back thirty years, and colour television was only just becoming widespread, the CD had not yet been invented, and there were no public mobile phone services at all. The role of digital technology in our lives is still changing, and more rapidly than ever. So it may come as some surprise to discover that the technology itself is not changing as quickly as you might imagine.

The technologies underlying the gadgets of the near future – the next ten years at least – will be the same technologies as those in use today. So in Chapter 1, I will survey today's gadgets in a bit more detail and explain how they work, so that we can understand the developments more fully as we peer into the future.

1 **Objects of Desire**

Apple's iPhone is an icon of the modern consumer electronics industry. It created a frenzy when it was released in the USA in July 2007, and in other parts of the world later in the same year. The iPhone is a mobile phone, but it can also connect to wireless computer networks; it is a music and video player, an atlas, a digital camera, a photograph album, a web browser, a calendar and an address book. There is only one button – all the other controls are accessible via a very responsive and, most people would agree, easy-to-use touch screen. That screen has 'multitouch' technology, with which the user can pinch or squeeze photos and web pages using two fingers, to zoom in and out. The iPhone was a success even before its release, and it quickly became an object of desire. To many people, it seemed like a real leap forward in technology – as if nothing else like it had ever existed. While there was genuine innovation in the design of this product – along with clever marketing and exquisite styling – there was no great technological leap forward involved in its development. It is a multi-featured 'smartphone' – and many other such devices already existed, albeit without quite the visual impact nor, many people would say, the same ease of use.

The point is that although the face of consumer electronics changes quite rapidly, the underlying technologies develop much more gradually. Processor speeds increase, but the processors themselves are still based on integrated circuits

Apple's iPhone – a digital object of desire.

COURTESY OF APPLE

made from semiconductors. Data storage and displays become cheaper and better, but they are still hard disks, solid-state memory and LCDs or plasma screens. And services such as wireless and mobile Internet access roll out more extensively and with enhanced speeds, but they are still based on digital information encoded in radio signals. All of this rapid development leads to generations of devices that are better equipped and more widely available than previous ones. Specific features and styling are hard to predict in new products, but for years to come, all new consumer electronics products will be built around the same core set of technologies.

Although the face of consumer electronics changes quite rapidly, the underlying technologies develop much more gradually

The iPhone will not be such big news two years down the line, and may well have been replaced in many consumers' minds, if not their pockets, by products that are equally innovative. But those new products will still work in basically the same way. And so, in constructing a vision of tomorrow's world of consumer electronics, we set out by deconstructing today's.

There is a huge range of items that could be tagged as 'consumer electronics'. They include simple children's toys, quartz clocks, watches and stopwatches, battery chargers, novelty doorbells, electronic calculators, radio-controlled model cars and even musical birthday candles. Also, appliances such as fridges, food mixers and washing machines increasingly rely, in part at least, upon electronics. Usually, however, the definition of consumer electronics is much narrower: it is normally restricted to digital devices that help us communicate, that keep us entertained and that give us access to information. That still encompasses a huge range of products, including desktop and portable computers, mobile phones, thin, flat-screen televisions, e-book readers, portable navigation units, small robots, personal media players, digital radios, digital picture frames, computer printers and scanners, digital cameras and camcorders, set-top television boxes, satellite receivers, personal video recorders, projectors and games consoles. All things I desire.

The kind of desirable items listed above is sometimes described as 'brown goods' – because in the early days of the consumer electronics industry, radios, telephones and televisions were commonly boxed in brown Bakelite, an early plastic. Modern, hi-tech brown goods are available in the high street, in shopping centres and, of course online. They are normally grey or white or cased in sleek brushed metal – rarely brown. Whatever colour they are, they are extremely popular: in 2007, consumers worldwide spent more than £165 billion ($340 billion) on these gadgets (note: in this book, as in most books, magazines, websites and news reports nowadays, one billion equals one thousand million – even in Britain – and one billionth equals one thousand-millionth). In that year, sales of mp3 players and digital cameras rose by 20% – as they had been doing for the past five years – and sales of LCD and plasma screens were around 50% more than in 2006. Sales of mobile phones reached 1.2 billion – again, an increase of around 20% on the previous year.

Whatever you call them, **however** you classify them – and **whatever their reason for being** – by buying into these 'must have' **gadgets**, we are **dramatically changing** the way we socialise, do **business** and organise our lives

Playing the Game

One of the devices that best illustrates the rapid advance of consumer electronics, and its increasing influence on the lives of millions of people worldwide, is the games console. The two most sophisticated games consoles have enormous computing power, thanks to their 'multi-core' processors. At the heart of Microsoft's Xbox 360 is a triple-core processor – basically, three processors working simultaneously. Sony's PlayStation 3 has a 'Cell' processor, which has a single processing core, but with six separate processing elements built-in, all working in parallel with the main core.

The feature of video games that has changed most noticeably in the past decade is the quality of the on-screen graphics. The most powerful consoles can now produce realistic output at the quality and resolution of high-definition television. Inside a games console, a powerful graphics processor prepares the signal for output to a display. The graphics processor carries out a significant amount of the processing power required for games – often including working out how objects will look from different viewpoints as the player moves around in the game. Just as important as what the game looks like is how it feels to play. In Assassin's Creed, released in November 2007, the game designers endowed incidental (non-player) characters with individuality and realistic behaviour. Players can explore three huge virtual cities, each rendered in exquisite detail. Inside the game environment, the player can jump from building to building, gripping anything that protrudes from any building.

In some countries, gaming is so much part of people's lives that it has become a popular spectator sport. In South Korea, two television channels are dedicated to video games, and star players release DVDs of their best games. In the countries where gaming is most popular – South Korea and Japan – professional gamers have huge followings. In 2005, a hundred thousand people turned out to watch the progress of South Korean pro-gamer Lim Yo Hwan as he played his favourite game.

The two most powerful games consoles on the market are Microsoft's Xbox 360 and Sony's PlayStation 3. Both are so-called 'seventh generation' machines – which gives you an idea of how long the leapfrog game of technological progress has been going on in the computer gaming industry.

Whatever you call them, however you classify them – and whatever their reason for being – by buying into these 'must have' gadgets, we are dramatically changing the way we socialise, do business and organise our lives. So what are they, really? And what do they have in common? If we are to understand how these gadgets might evolve, we need to know something about how today's versions work. So if you are the kind of person who is familiar with terms such as RAM and gigahertz, but you don't really know how it all works, then the following should get you up to speed. It will get a little bit involved, but I think it's important that people have an idea of the magic that goes on inside their treasured gadgets. After all, love them or hate them, desire them or not, these devices work hard for you every second they are switched on.

Computers Everywhere

You might think of your gadgets in two categories: computers and the rest. But what makes a computer a computer and other devices just, well, other devices? You may be surprised to find out that there is little difference – that all modern consumer electronics devices are, in fact, computers. That simple mp3 player? A computer. That digital camera? Likewise. Even a digital television is a computer.

A computer is defined by its 'architecture' – the existence and organisation of the main components inside. In particular, every computer has a central processing unit (a processor) to carry out sets of instructions and perform calculations, working memory to store those instructions and the interim results of calculations, longer-term storage, and some kind of input and output.

This is why, almost without exception, all modern consumer electronics products are computers. Every set-top box, every wireless router, every printer, scanner and even every CD and DVD player is a computer. Many of these devices are dedicated to carrying out a limited set of functions. All of those functions are realised by a single program called the firmware. This program loads when you switch on the device.

To reinforce this idea that every digital device is a computer, consider the fact that you can reprogram many of what you might think of as dedicated devices to do other things. For example, hobbyist hackers can install their own firmware and software in their games consoles, media players and even digital cameras. In 2007, an astrophysicist at the University of Massachusetts, USA, loaded a different operating system onto six Sony PlayStation 3 video game consoles, then linked them together to make a single, extremely powerful supercomputer.

Electric currents rush around inside chips like these, hidden from view inside your digital gadgets, making magic happen.

Dedicated Devices

Electronic products that have limited, specific functions – rather than the ability to load and run a variety of programs as personal computers do – are called dedicated or embedded devices. They are often furnished with a specially designed and custom-built chip called an application-specific integrated circuit (ASIC). There is an ASIC inside your digital camera, in your broadband modem and in your wireless router. An ASIC is a complex circuit specially designed and custom-built to enable just the functionality of each device. Having an all-in-one solution like this keeps power consumption low and cost down and improves performance and reliability. Most gadgets also contain another type of integrated circuit, with the immediately forgettable name of 'application-specific standard product' (ASSP). These circuits carry out specific functions, such as processing sound or video or producing a radio frequency signal for a wireless device but no more. So this is a 'modular' approach to electronic product design, since devices can be built with a number of 'off the shelf' ASSP chips, each with a different function.

Solder a few ASSPs and an ASIC onto a circuit board together with a memory chip and load a firmware program into the memory chip; connect the whole thing to a power supply and a large LCD screen – and a video source – and box the lot in a steel and plastic casing, and you have a high-definition television. Obviously, it is a lot more complicated than that, but that is the basic recipe. There are ASICs and ASSPs at the heart of mobile phones, in personal video recorders, network routers, and set-top boxes, and in most other dedicated devices.

The personal computer (PC) lies at the core of our digital existence – as it has been for increasing numbers of people since the early 1980s. It is a general purpose device. At its heart is a microprocessor – a more flexible and more powerful integrated circuit than you would find in a device dedicated to a limited set of functions. The microprocessor enables a computer to load and run any number of different programs. In 2007, more than one hundred million desktop, laptop, notebook and 'ultra-portable' personal computers were sold to consumers worldwide. Another 150 million went to business customers. Modern mobile phones can do a lot more than just make and receive phone calls and send and receive text messages. For that reason, they really are closer to what we would think of as computers. They too have a microprocessor.

The Numbers Represent

We've all heard about the Digital Revolution. Today's consumer electronics products – general-purpose PCs and smart phones as well as dedicated gadgets – are all digital devices. And 'digital' is the future. The New York journalist Walt Mossberg has said that 'We are in the path of a digital tidal wave.' So what does 'digital' really mean? It's simply that our gadgets represent information and programs as numbers. The microprocessor at the heart of a computer is a number cruncher: all it does is manipulate numbers, albeit extremely rapidly. As processors become faster – able to do more number-manipulations per second – they are able to process digital information more rapidly. And that digital 'information' is typically composed of images, video, sound, presentations and programs. So, as consumer electronics move ahead, we are more easily able to store and manipulate this kind of information – for convenience, information and entertainment. So just how is it possible to represent pictures, sound and video as numbers?

Consider digitising an image. The simplest way is to break down an image into a large number of dots or squares. Then, each of these picture elements, or pixels, can be assigned numbers based on the brightness and colour at that point in the image. The collection of all these numbers will successfully represent the image. A computer can use the numbers to reconstruct the image on a monitor or print it. The more pixels there are, and the more numbers are used per pixel, the better the detail in the digitised image. This is why good-quality digital images take up relatively large amounts of storage space on hard disks or memory cards. It is also why digital cameras have sensors that can measure brightness and colour at millions of points in an image.

It is easy to extrapolate from digitising individual images to digitising video, which is composed of a sequence of digitised still images, or frames. In practice, there are different schemes by which the numbers are arranged in the files that represent digital images and digital video. For example, there is 'meta-data' – information relating to the camera make and model, the program used to manipulate it, the date it was created and the image size. But the principle is straightforward.

The New York journalist Walt Mossberg has said that **'We are in the path** of a **digital tidal wave'**

Audio – sound signals – can be represented numerically in a similar way. Sound is caused by variations in air pressure, which spread out in all directions as waves. If a microphone is near to a source of sound, it produces a varying electric voltage that matches the sound waves. In fact, the variations in voltage form a direct copy,

Digitising an image – break the image up into tiny picture elements, or pixels, and assign a number to each depending on its colour. The more pixels and the more numbers used, the better the representation of the image will be.

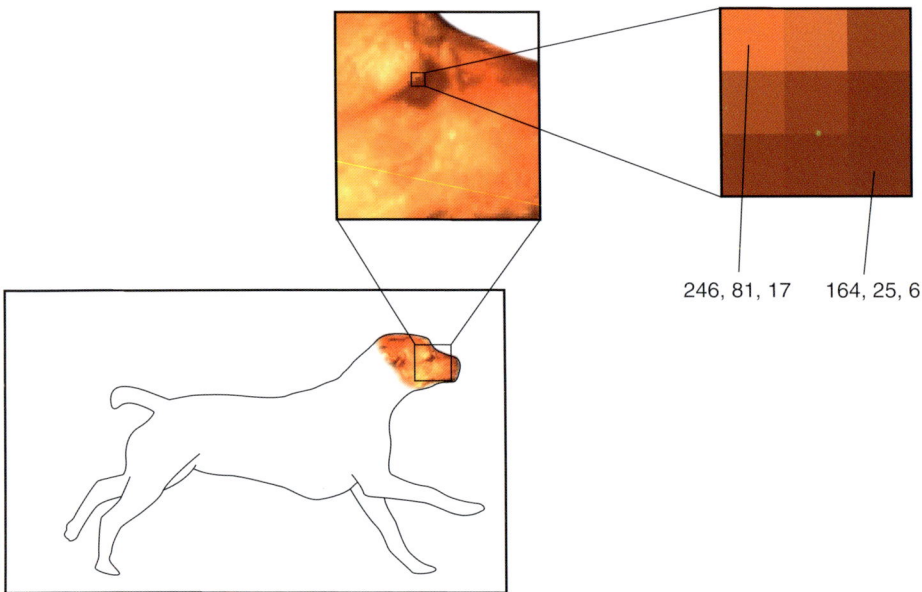

246, 81, 17   164, 25, 6

or analogue, of the sound wave. An electronic circuit called an analogue-to-digital converter measures, or 'samples', the voltage thousands of times each second. So, digital sound is simply a stream of numbers that represent the air pressure in the original sound. If you drew a graph using those numbers, it would look exactly like the original wave. Again, the more samples (per second in this case), the more faithful the digital representation of the original sound. And once again, there are different formats and clever ways of working with those streams of numbers, but the principle is clear and simple. Digital devices can manipulate and store those numbers. Using the right sets of instructions, a processor can even

Just behind the lens of a mobile phone's camera lies an image sensor like this one. It is made up of millions of individual sensing elements, each of which produces an electrical voltage that corresponds to how much light falls on it.

generate new sounds never heard, just by producing the relevant streams of numbers. This is the principle behind digital synthesisers, and most computer sound cards have synthesiser circuits on-board that are used to create sounds in some games or any programs where sound effects or electronically produced sounds are needed.

When you **download a web page**, or **pictures** and sounds from the **Internet**, you are receiving **millions** of individual numbers

Digitising sound – A microphone captures a sound wave as a varying audio signal. A digital device samples the level of the audio signal thousands of times every second. The more often the audio signal is 'sampled', the more faithful the representation and the clearer the sound will be.

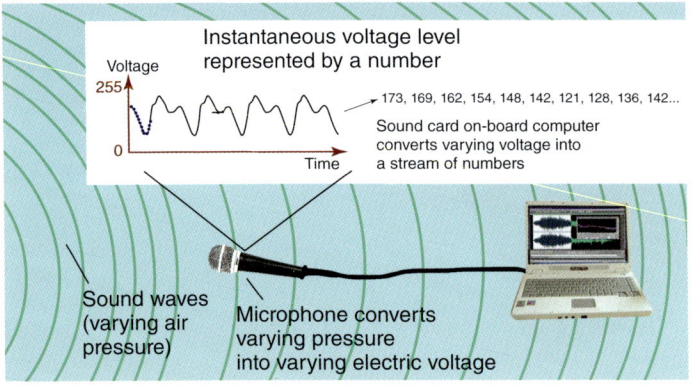

Instantaneous voltage level represented by a number

Voltage
255

173, 169, 162, 154, 148, 142, 121, 128, 136, 142...

Sound card on-board computer converts varying voltage into a stream of numbers

0

Time

Sound waves (varying air pressure)

Microphone converts varying pressure into varying electric voltage

Alphabetic characters provide the simplest example of digitisation. There is a simple one-to-one coding scheme: each letter has a unique number to represent it. On most computers, upper case 'A' is represented by the number 65, for example. There are also codes for program instructions, as well as numeric characters, symbols and punctuation marks. Again, many different coding schemes exist, but every document or program stored on a computer really is simply a large collection of numbers. When you download a web page, or pictures and sounds from the Internet, you are receiving millions of individual numbers. This is why broadband Internet connections are opening up so many possibilities – when you can download millions of numbers every second, you can stream video and sound, while browsing web pages, for example.

All in Bits

If you could see the numbers that represent images, video, text or other characters written down as the computer 'sees' them, you would see endless arrays of ones and zeroes. Computers represent numbers using the binary system – also called 'base two' – while we humans tend to use the decimal system, or base ten. The binary system has only two available digits – 0 and 1; we use ten digits: 0, 1, 2, 3, 4, 5, 6, 7, 8 and 9. Inside a computer, the two binary digits, or 'bits', are commonly represented by voltages and electric currents, or by magnetic fields: 0 and 1 might be 'on' and 'off', 'high' and 'low', or 'magnetised this way' and 'magnetised that way'. In addition to the flexibility of being able to use any of these two-state systems to represent numbers, it is much easier to design circuits that can carry out arithmetic with just two digits than with ten. And, as we pointed out above, arithmetic is central to how computer processors work. It may seem amazing, but all your videos, images, text documents and software would be indistinguishable to the untrained eye if they were written out in their true binary form. They are all just large collections of ones and zeros.

Do you know the difference between a bit and a byte? Basically, it's a factor of eight. Why? The first personal computers used groups of eight bits to represent letters, numbers and other characters. Using eight bits, there are 256 different numbers available: from 00000000 (zero) to 11111111 (255). So, it was natural to measure amounts of information represented by binary numbers with a special unit – the byte, equal to eight bits. Nowadays, most personal computers – and other consumer electronics devices – are based on 32-bit groups; some on 64-bit groups. Still, the byte remains the dominant unit. One thousand bytes make a kilobyte (kB); one million bytes make a megabyte (MB); and one billion bytes make a gigabyte (GB). These terms are all familiar to anyone who uses the Internet, as are 'kilobits per second' – thousands of binary digits received or sent on a network each second. Megabits per second and

When you hold down the shift key on a computer keyboard and press the key marked 'A', the keyboard will send a stream of electrical pulses to the computer processor that represent the number 65 (in binary notation, 01000001).

gigabits per second are millions and billions of bits sent or received per second, respectively. All of these familiar units simply measure quantities of numbers. One kilobyte is simply one thousand eight-bit numbers – or even more simply, eight thousand 0s and 1s. Many consumers have only just got used to the term 'gigabyte' – since hard disk capacities used to be measured in megabytes. In the next few years, you will increasingly be hearing the term 'terabyte' – one terabyte is a thousand gigabytes.

To give you a bit more insight into what really goes on inside your digital devices, consider how many binary numbers are needed to represent one second of sound at 'CD quality'. Every second of CD-quality sound is represented by 44,100 individual samples in each of the two stereo channels. Each sample is represented by a 16-bit binary number. So, when you record or play back CD-quality sound on a media player, a large number of binary digits must be processed every second to keep up. That number is 44,100 (samples) multiplied by 16 (bits per sample) multiplied by 2 (for stereo) – a total of 1,411,200 binary digits. This is equivalent to just over 1,411 kilobits, or slightly more than 176 kB (1,411 divided by 8, the number of bits per byte).

When sound is stored as an 'MP3' or other compressed format, near-CD-quality sound reproduction can be achieved with far fewer bits per second. This is done by altering the stream of numbers using clever mathematical functions called 'compression algorithms'. These algorithms are based on eliminating the bits corresponding to parts of the music that are too quiet or too high-pitched for our ears to hear well. A typical mp3 track downloaded from an online music outlet will be encoded at 192 kilobits per second – less than one-seventh the rate of the 'raw' CD-quality sound. As a result, a three-minute track stored on a digital music player will take up just over 4 MB of memory, while the same track at raw CD-quality will require nearly 32 MB. You notice this difference if you play an mp3 over a high-quality system or very loudly, but the difference is negligible most of the time, and those of us who buy music online would much rather have our music delivered swiftly and in acceptable quality than to wait for much longer for a better quality track. And it means we can have a capacity of many thousands of tracks on our media players, rather than a few hundred – if we so desire.

In the same way, we can work out how many bytes are needed to store a digital image. You can, of course, scan a photograph line by line, the scanner sending the relevant streams of numbers to a computer. But nowadays, most images are produced in digital form at source – inside a camera. Behind a camera's lens, images fall on a light-sensitive device – normally a charge-coupled device, CCD, which is a semiconductor chip similar to a processor or a memory chip. A typical CCD has millions of sensor elements on its surface, in a grid pattern. A CCD like mine with a grid 3,264 wide by 2,448 high has a total of nearly 8 million sensors – and would be said to have a resolution of 8 megapixels. Each tiny sensor produces a voltage that depends upon the intensity of light falling on it. These voltages are digitised – represented by binary numbers – normally using 8 bits for each pixel. In most cameras, there is a coloured filter array on top of the CCD, which consists of red, green and blue filters in a regular arrangement. The result is that a camera produces three images – one red, one green and one blue – which produce a full-colour image when recombined. The processor inside the camera combines all the information to produce a 24-bit number for each sensor.

So, a 'raw' 24-bit per pixel image produced by an 8-megapixel digital camera (with 8 million of those individual sensors) requires a total of 192 million bits (24×8 million), or 24 MB. As with sound, there are various compression schemes by which digital images can be represented using far fewer bits than this without noticeably losing image quality. The best known is JPEG – the acronym stands for 'Joint Picture Expert Group', the organisation that defined the format. Most digital cameras store images as JPEGs; in the case of an 8-megapixel camera, each JPEG image requires about 5 MB rather than 24. This kind of compression is most important of all in digital video – especially in portable devices, where storage may be a premium. One second of uncompressed, full high-definition (HD) digital video, with sound, requires nearly 1.5 gigabits per second. That is 15 billion 1s and 0s every second.

The fact that digital devices represent information using binary numbers is very important. It allows real flexibility in the way those devices store and process information

The fact that digital devices represent information using binary numbers is very important. It allows real flexibility in the way those devices store and process information. The message is independent of the medium. So, binary-coded information can be represented by laser light in a CD burner or in an optical fibre or by radio waves in a wireless network. They can be represented by electric voltages and currents inside integrated circuits and along cables and wires. Or they can be represented by magnetic fields in hard disks. And in this digital age, we are constantly surrounded by information in all these forms. This also makes it possible for different digital electronic devices to communicate with each other. So, for example, an image produced by a digital camera, stored as numbers on a memory chip using magnetism, can be sent as numbers encoded in radio signals, to a network router, which might send electric pulses along a wire, ultimately reaching a processor in your computer. You can back up the image as billions of tiny depressions on a spiral track in the surface of a CD or DVD. And because information can be represented in any medium that can represent numbers, this 'digital paradigm' is also future-proof. Information is represented by magnetisation on today's hard disks; high-capacity storage in the future will probably be quite different. Similarly, computer processors are currently manufactured from silicon or other semiconductors, but that need not be true in the future. Again, the medium is not the message.

Storing Bits

We need somewhere to put all these bits and bytes; an essential function of any consumer electronic gadget is the ability to store digital information. There are two basic types of storage: working memory and long-term memory. The working memory is normally in the form of several integrated circuits carrying RAM (random access memory). It is important that the processor can access any particular piece of information in the RAM very quickly indeed. When a device is working, large amounts of digital information are continuously exchanged between the RAM and the device's processor. RAM is described as volatile: this simply means that information stored in it is lost when the device is switched off.

The firmware – that basic program that controls how a device operates – is normally stored in a different kind of chip, called 'read-only memory' (ROM). This is an integrated circuit soldered to the device's main circuit board. The information in a ROM chip cannot be overwritten, except under special circumstances, such as when the firmware needs to be updated. It is copied to the working memory at start-up, but remains unchanged on the chip, ready for the next time the device starts up.

So ROM is long-term storage but it cannot be overwritten, and RAM can be overwritten but is not long-term storage. Many devices also have long-term storage in which the information can be overwritten. Even a dedicated device such as a television has some way of backing up information, such as channel favourites, brightness and contrast, and the picture's aspect ratio. And of course, a media player can store digital representations of hundreds of songs, and you can erase tracks and add new ones. And a mobile phone has somewhere to store a list of names and numbers, along with text messages received and the details of the service provider. In these kinds of devices, the storage medium of choice is often 'flash memory'.

Flash memory is a solid-state device – it has no moving parts – so it is ideal for devices that may be subject to vibration or repeated movement. The name comes from the fact that when erasing information, large portions of the available memory in a flash chip are wiped, or 'flashed' at once. It is very light, too, and it takes up little space and retrieves stored information very quickly. All of this makes it ideal for devices where light weight and robustness are important: in mp3 players that you might listen to while running or working out at the gym; in mobile phones, which may be dropped, thrown into bags and pockets, and which must be very portable. However, flash memory tends to be prohibitively expensive to incorporate much more than a few gigabytes of storage into a single device. Some PCs – in particular laptop, notebook and subnotebook PCs – have flash-based storage. This reduces the drain on battery power, and makes for a more robust and lighter machine – but for now, it does add significantly to the cost of the computer. Those USB memory sticks you carry around contain flash memory chips, as do the memory cards you can buy for your digital camera.

Of course, most PCs – and many other devices – use hard disks for long-term storage. Even a cheap hard disk drive can hold vast amounts of digital data. Until flash memory chips become cheaper, all media players that have more than a few gigabytes of storage are nearly all hard disk-based. The smallest hard disk drives are small enough to fit inside a small media player, but can have hundreds of gigabytes of storage space. Inside ROM, RAM and flash memory chips, binary digits are stored as electric charges or electric fields. But inside a hard disk drive,

The highly polished surface of a hard disk platter, with the read-write head poised just above it.

binary digits are stored using magnetism. A hard disk drive contains set of stacked discs that rotate together at high speed – typically 7,200 revolutions per minute (that's 120 times every second). Perilously close to both sides of each rapidly spinning disc is a read/write head, with a tiny electromagnet at its end. The head's electromagnets rapidly magnetise tiny regions of the discs' surfaces in one direction or the opposite, to represent the binary digits 0 and 1. Each read/write head is attached to the end of a pivoting arm that can sweep across the disc, so that it can position the read/write head next to any part of the surface of the rotating disc. Clearly this is a high-precision process.

Sending **information** from **one digital device** to another is a **matter of transferring streams of binary numbers** between the two devices

Making the Connection

Your consumer electronics gadgets would not be much use if they could not communicate with other such devices. Communication is vital in more subtle ways than you might immediately imagine – it's not just about phone calls, text messages and emails. Yes, your mobile phone must be able to communicate with a base station. But transferring music and video from your PC to a media player is also an example of one device communicating with another. And a PC must be able to send information to printers, and to and from other computers via the Internet or local networks. Electronic musical instruments may need to communicate digital information about note pitch, duration and volume to multitrack recording software. Even normally disconnected devices may need their firmware updating from time to time. Sending information from one digital device to another is a matter of transferring streams of binary numbers between the two devices. The rate at which data pass along a connection is called the bandwidth, and is measured in terms of how many bits or bytes can be transferred each second.

The most obvious method of connecting two devices is to use a cable, and there is a wide variety of different ones available. The most common is the USB cable. A host of different devices, including digital cameras, printers and scanners, media players, and mobile phones can connect to each other – or more normally to a PC – via a USB cable. The current standard, USB 2.0, allows data transfer at up to 480 megabits per second (60 MB per second). At this speed, that 4 MB mp3 music track described above would shoot along the cable in a fraction of a second. USB has the added benefit that it can charge the battery in a portable device while information is being transferred. Another common wired connection is the Ethernet cable – it looks like a slightly plump telephone cable, with plugs similar to those on a telephone cable at either end, too. Originally designed for connecting PCs together to form networks, Ethernet cables can also carry information to and from attached storage devices, video games consoles, printers, and network hubs and routers. The current fastest Ethernet technology, called gigabit Ethernet, transfers information at a rate of up to one gigabit per second – although a fibre-optic version of the Ethernet cable can achieve speeds ten times that.

It is far more convenient to send information through the air, so that you don't have to plug one end of a cable into each device. Like wired connections, wireless connections also take many forms. For example, information can be coded into invisible infrared radiation and passed from one mobile phone to another – this is actually not much more complicated than what happens when you point your remote control at the television. It's just infrared LEDs flashing on and off in patterns not unlike Morse code.

More commonly, mobile phones use Bluetooth – a short-range system that codes information into high-frequency radio waves. Most people are familiar with this technology because their mobile phones are equipped with it. Wireless hi-fidelity headphones and remote controls for media players are also becoming popular – headphone wires can be really annoying. The most recent version of Bluetooth, adopted in August 2007, may widen the uptake of this technology, because it has made the process of connecting two devices simpler and more secure. However, Bluetooth bandwidth is currently limited to 2.1 megabits per second. The next Bluetooth standard should increase this to 480 megabits per second, bringing it in line with the speed of USB.

Bluetooth tends to be used for 'personal area networks' because of its limited range. The other familiar wireless connection, Wi-Fi, is normally used to connect devices together to form a larger wireless network. A Wi-Fi connection is the wireless equivalent of an Ethernet cable. It is very common in home networks and, of course, in 'wireless hotspots'. Just as with Ethernet, you can connect directly to suitably equipped printers, and share Internet connections between computers. But you can also, simultaneously, connect to some mobile phones and media players. Many consumers' experiences of Wi-Fi are not entirely positive, however. It can be unreliable – subject to interference or simply annoyingly difficult to administer if the set-up does not go smoothly. Sound familiar? A similar technology, called WIMAX, is available in some areas. It has a much higher bandwidth, and a greater range, than Wi-Fi, and it is very reliable, making it suitable for making wireless voice calls without using mobile phone network. Several cities have WiMax Internet connections available for subscribers – but you need a WiMax-enabled device to use it.

Observing the Protocol

There are two main modes of communication between connected devices. These are 'circuit switching' and 'packet switching' – each has a different way of delivering the information. In a circuit-switched connection, a dedicated channel is set up between the connected devices. This approach originated in the first telephone exchanges, where operators physically connected two callers' lines, forming a dedicated circuit that was available for the duration of a telephone conversation. This guarantees swift and consistent transfer of digital information. Mobile phone networks employ circuit switching to carry voice calls. However, when a channel is dedicated to a particular pair of devices, no other devices can use the medium, and available bandwidth can be effectively wasted when, for example, the connection is not actively in use. This is not a problem if you are connecting your digital camera to your own computer via a cable, but in a computer network, it can be. For example, if someone were to send a large file across a network, other users' computers would have to wait until the transmission was finished before they could use the network.

The packet-switched approach involves breaking up digital information into small nuggets called packets. A packet is typically a few hundred bytes in size. A digital image sent from one PC to another on a packet-switched network is broken up into dozens or hundreds of packets by the originating computer. While they make their way across the network, these packets will be interspersed among other network traffic. Every packet might take a different route from all the others. Once all of the packets have completed the journey, the destination PC reassembles them to form an identical copy of the original file. When the network is busy, the packets take slightly longer to reach their destination, but the advantage is that the connection is always available.

Each packet on a packet-switched network carries extra information that allows it to be routed correctly to its destination. Exactly what extra information is carried on a packet depends upon what particular 'protocol' is being used, but it always includes the network address of the originating device and an address of the destination device, as well as the time it was sent. Most packet-switched networks use the Internet protocol (IP) to manage the addresses of connected devices. Every device connected to an IP network has an IP address. The Internet is a vast, global IP network of IP networks – an 'internetwork'. And every device connected to the Internet has a unique IP address.

Although mobile phone networks use circuit switching for voice calls, they use packet switching when they transfer data – for example, when a user wants to send and retrieve emails and multimedia messages, browse the Web or access information services made available by their network provider. The main advantage to users is that they are charged according to how much information they transfer, not the duration of the connection. Ethernet and Wi-Fi connections also use packet switching, while Bluetooth can operate in both packet-switched and circuit-switched modes.

That concludes our whistle-stop tour of the principles behind today's digital technologies. If you have followed these explanations, then you are more than ready for our journey into the future – both near and far, since digital, computer-based technology will continue to be the norm for many years to come.

The Future in Your Hands

Being 'connected' is of particular importance for portable devices. In a world in which so many of us are constantly on the move, portable devices offer real convenience. We love our pocket-sized devices, but there are drawbacks. For example, the more features a device has, the greater is the drain on a small battery. And the smaller and more feature-packed the device, the bigger the problem. It is also difficult to enter text on tiny keyboards – especially if you are used to full-sized computer keyboards and you need to type large amounts of text quickly. And then there is screen size: we are used to large screens at home, both with desktop computers and televisions. Magnifying small portions of a web page or a document is never quite the same as seeing the whole thing at once, and watching videos or looking at photographs on a small screen can be equally disappointing. One other problem is that, with so many different functions, the most useful portable gadgets are often the most difficult to use effectively. Finding your way around complicated menus can be frustrating for a non-technically minded consumer, who wants gadgets that 'just work'. Who wants to read through instruction manuals or search the Web for help?

Finding your way around **complicated menus** can be frustrating for a non-technically minded **consumer**, **who wants gadgets** that 'just work'

One solution to the problem of tiny keyboards, and to the problem of shrinking battery sizes, is simply to use a slightly larger device. A class of computer called subnotebooks lies between the notebook PCs and smaller, handheld devices – in both size and power. Subnotebooks are almost small enough to fit in a pocket, and they are very light, but they maintain most of the functionality of genuine notebooks. Most of these small computers have the ability to connect to mobile phone networks as well as Wi-Fi networks, for browsing the Web as well as for making phone calls.

The mainstay of the portable gadget world is the mobile phone. In 2007, consumers worldwide bought more than a billion mobile phones. The total number of subscriptions to mobile services passed 3 billion for the first time in July of the same year; that's nearly one mobile for every two people on Earth. It took about twenty years for the first billion subscriptions, two and a half years for the second billion, and less than two years for the third. The number of people using mobile phones is actually just over 2 billion, because a significant number of people have more than one subscription.

Mobile phones are popular across all ages; a survey carried out at the end of 2006 found that around 90% of children aged 12 in Britain own a mobile phone. In 2007, according to the UK's Mobile Data Association, an average of 5 million texts were sent every hour in Britain – a total of around 50 billion for the year. Most phones sold in the past couple of years have built-in cameras; increasingly, these cameras have picture quality that rivals that of expensive stand-alone digital cameras. Worldwide in 2006, consumers bought 500 million camera phones, and a 2007 study by Strategy Analytics predicted that one-third of the global population will own a camera phone by 2011. These numbers are staggering. In countries with the highest mobile phone use, and with the most advanced mobile infrastructure, mobile television services are already thriving, and millions of people also use their phones as a form of payment, taking the place of credit cards.

Since 2004, mobile phone users in Japan have been able to use a facility called Osaifu-Keitai (literally 'mobile wallet'). This is a 'contactless' payment system that can act as cash, a credit card, and a direct ticket payment and booking system for transport and concert tickets. To use the system, customers simply place their phones near to a special reader, information passes between phone and reader via short-range radio signals, and the transaction is carried out in seconds. This system is being trialled in several other countries.

Smartphones

So-called 'smartphones' typically incorporate the ability to read and send email, browse the Web, download and run new software, and play games. These devices are basically handheld general-purpose computers without the processing power or storage capability of a desktop or portable personal computer.

Smartphones bring real convenience, especially when they include many different functions within a single device that fits in your pocket. Being able to make or take a mobile phone call without removing the earphones you have been using to listen to your favourite music tracks, while also looking at your diary or address book, is a real bonus. Add to that the ability to carry precious memories as photographs and videos, and to access and even contribute to a world of information via the Internet, and it is no wonder portable devices are so popular and desirable. There are other handheld, pocket-sized devices that are not mobile phones at all, but which offer all of the other functions present on a smartphone. In addition to carrying storage for music, digital photographs and video, and documents, these gadgets are typically able to connect to wireless networks, enabling users to check email, browse the Web, and download music or video.

Nokia's popular N95 smartphone, along with the equally popular Blackberry smartphone, by Research in Motion.

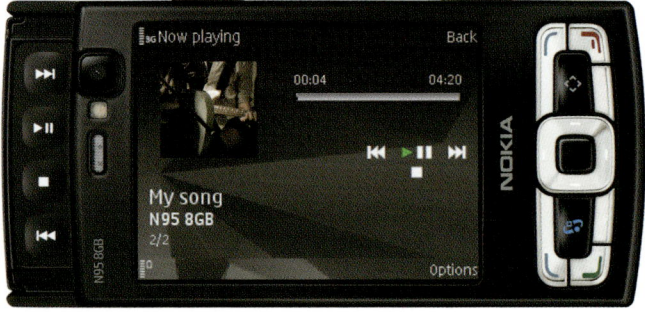

Rapid uptake of **mobile phones in poorer nations** seems to be part of the reason why consumer **electronics** industry surveys often refer to such nations as '**emerging markets**' **rather than** 'developing countries'

Mobile phones are not only changing people's lives in the rich nations. They are one of the few consumer electronics devices to have a truly global impact. Rapid uptake of mobile phones in poorer nations seems to be part of the reason why consumer electronics industry surveys often refer to such nations as 'emerging markets' rather than 'developing countries'. You might see this as a positive development, since the term 'developing countries' can be seen as patronising. Alternatively, you might think it is a reflection of a cynical, profiteering industry that will go all out for profits wherever it can find them. After all, do people in poorer nations truly need to feel they should buy into our obsession with digital technology?

Well, there is an argument that electronics really can be beneficial to people. Mobile phones can be an effective tool in the fight against poverty. For example, they enable farmers in remote areas to check out current market prices for their produce. One scheme, run by the Senegalese company Manobi, already provides a service to tens of thousands of farmers, allowing them to trade directly via their phones. Camera-equipped mobile phones allow farmers to make and share tips or techniques via video clips, or warn of the spread of disease. Other workers use their mobile phones to find work in nearby towns and villages; people living far from doctors receive basic medical diagnosis and advice from doctors without having to travel tens of miles to the nearest surgery. Is this digital technology making a real positive difference to people's lives, or is it a slippery slope towards a world in which everyone has to remember to charge their mobile phones and where people pay a significant amount of their income just to keep up with the technology? Mobile phones and airtime cost a higher proportion of a poor person's income than a rich person's.

Many people in poor and remote areas have begun using camera phones and smartphones in the same ways as people in richer nations use their PCs. Jan Chipchase, who carried out research into mobile use for Nokia in several countries, has said 'If you ask people what the three most important things they carry are – across cultures, and across genders and across contexts – most people will say keys, money and, if they own one, mobile phone.' The continent with the fastest growth in mobile phone use is Africa. In 2001, there were about 25 million mobile phone users across the continent; by 2007, there were around 200 million. Around 65% of Africans live in areas covered by mobile phone signals – mobile phone services require much less infrastructure than traditional landline connections.

Mobile phone booth in South Africa.

On Display

Portable or not, most of our devices need some way of displaying information. The most basic display technologies are simply indicator lights, such as those annoying ones that flash to indicate hard disk activity or the fact that information is passing to and from the Internet via your router. These are normally LEDs (light-emitting diodes), which are great because they use very little electrical power. Those simple numerical displays you still find on many DVD players use LEDs, too. A variation on the ordinary LED is the organic light-emitting diode (OLED), which is already being used in a wide range of devices, including illuminated keyboards and mobile phone keys.

OLEDs can be used to make screens that can display images and video as well as just letters and numbers on a keyboard. Some mobile phones and some personal media players have small OLED screens, and in December 2007, Sony released the first televisions to use OLED screens. The future appears bright for this nascent technology – we'll gaze forwards in Chapter 4. Most personal media players and mobile phones use a more familiar and much more established technology: the liquid crystal display (LCD). And at present, LCDs dominate the television market, too – along with plasma display panels (PDPs) – and they are likely to dominate for another five years at least.

There are no physical keys on this concept keyboard designed by Art. Lebedev, so the keys can be of any shape and size and colour – thanks to an OLED surface.

LCDs and PDPs have begun the long job of displacing cathode ray tube (CRT) displays – the heavy and bulky glass tube at the heart of traditional television sets and computer monitors. Worldwide in 2007, consumers spent around $95 billion (£50 billion) on PDP and LCD televisions in 2007. Part of the popularity of LCDs and PDPs is due to the introduction of high-definition television, which has a resolution of either 1,920 columns × 1,080 rows (more than 2 million pixels) or 1,280 × 720 (nearly 1 million). A standard-definition TV picture in the UK has a resolution of 702 by 576 (just over 400,000 pixels). High-definition pictures demand a bigger screen. Because LCDs and PDPs are much thinner than CRTs, they can be incorporated into much bigger televisions without significantly adding to the weight. But even thinner displays will be needed ultimately, so that televisions can be bigger still – since there are still higher-definition television pictures to come. Even existing standards allow for pictures composed of 3,840 × 2,160 pixels (more than 8 million pixels).

A phosphor coating inside each cell of a plasma display panel produces red, green or blue light.

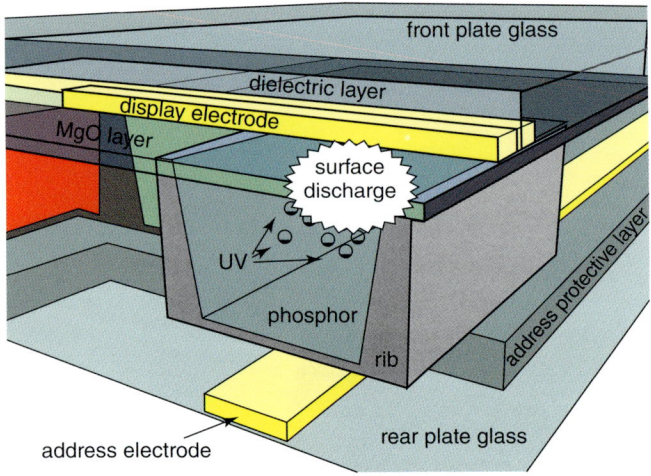

Digital Televisions

Each picture element (pixel) of a plasma display is composed of three subpixels – one that produces red light, one that produces green light and one that produces blue light. The human eye has only three types of colour-sensitive cell – one sensitive to red, one to green and one to blue. So by combining these three colours in the correct combination, three adjacent subpixels can produce the illusion of almost any colour. Each subpixel is a tiny, sealed cell.

The inside surface of each cells is coated with a phosphor – a substance that produces light when ultraviolet radiation strikes it. The ultraviolet radiation is produced when an electric field passes through an electrified gas, or plasma, which fills the cell. Each pixel is individually 'addressed' according to its distance 'along' and 'down' the panel. There are hundreds of thousands, or even millions, of cells – depending upon the screen's resolution – arranged as a grid that covers the whole screen. And plasma screens typically display a screen's worth of information 50 or 60 times every second.

Liquid crystal displays have been used in calculators, watches and many other devices with simple, small displays since the 1970s. An LCD consists of several layers – with glass at the front and a diffuse white light source at the back. The LCD makes use of a strange property of light called polarisation. There are two polarising filters, arranged at right angles to each other, so that they do not allow any light to pass through. Between them is a layer of liquid crystal: a liquid that consists of long molecules. An electric field passing through the liquid crystal causes the molecules to line up in an orderly fashion – like a crystal. As the molecules align, they affect the polarisation of light that passes through them. The result is that some of the light can now pass through the polarising filters. In a large LCD screen – as in a plasma screen – each pixel consists of a red, a green

An ultra-close-up shot of an LCD computer screen. From a normal distance, the background would appear white – because all the red, green and blue subpixels have the same intensity.

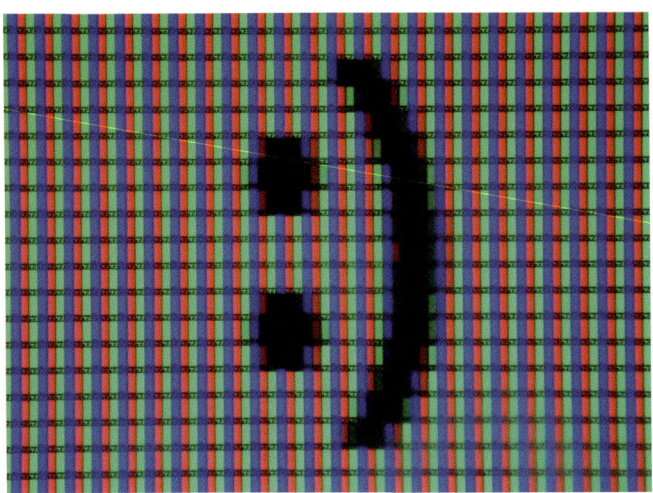

and a blue subpixel. Just behind the front glass is a chequered colour filter with red, green and blue squares. Each subpixel sits directly behind either a red square, a green square or a blue one. The pixels are arranged in a grid, and each subpixel addressed individually. Miniature versions of LCD screens also feature in LCD projectors: the backlight is replaced by a projector bulb that is bright enough to project an image of whatever is on the LCD screen onto a screen or a wall.

In this chapter, I have painted a rounded portrait of today's consumer electronics – but with a very broad brush. I have investigated only a fraction of the devices that are available, and only scratched the surface of even what consumer

How an LCD display works.

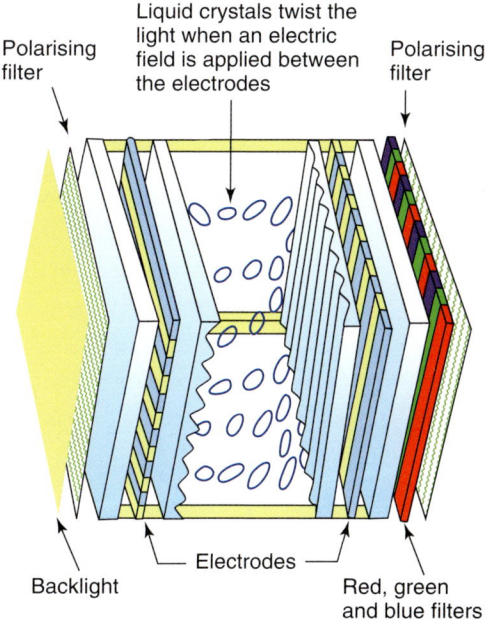

Polarising filter

Liquid crystals twist the light when an electric field is applied between the electrodes

Polarising filter

Backlight

Electrodes

Red, green and blue filters

electronics can already do. Science fiction author Arthur C. Clarke famously wrote 'Any significantly advanced technology is indistinguishable from magic.' In a sense, our gadgets can do magic: they make moving pictures of real people appear on glass screens; they let us talk to people far away while we walk in the park; the virtual worlds we create are magical playgrounds in which anything is possible. The magic powers of consumer electronics are scalable. As we develop ever-faster processors, better storage devices with more capacity, faster connections by which we can transfer ever more digital information, and display technologies that are even better than the ones we already have, the magic will grow and become more impressive. It all means that I will have to shell out even more of my hard-earned cash to satisfy my desires.

But how far will it all go? Where and when will it all end? And what kind of world will we be living in then? Will we all have cheap but immensely powerful and convenient house computers acting as intelligent personal computers? And if so, is that something you will want? Whatever happens, we consumers will play a vital role in marshalling the progress of these technologies – for the foreseeable future at least. But it is the consumer electronics industry that will be forging that progress. Designers, electronics engineers, physicists, materials scientists, software engineers, factory owners, factory workers, venture capitalists, sales and marketing executives, advertisers, retailers and service providers – all of these people and more are instrumental in the lucrative industry behind consumer electronics.

And so, in Chapter 2, I'll try to answer the following questions about the industry of our desires. How and where are our gadgets designed and manufactured? What are the environmental consequences of making and using these objects of desire? How can the men and women of this huge, disparate enterprise continue to make the gadgets we love ever more appealing? And in a business where competition is a vital part of sustaining progress, will it ever be possible for all our gadgets to be fully compatible with each other?

The Industry

Every consumer electronics device is the product of creative ideas and clever design; it has been fabricated ultimately from rocks and oil, assembled in huge factories, and brought to us by transport networks and retail outlets. Like the gadgets it produces, the consumer electronics industry is evolving all the time. Competition is fierce, and the only way to survive and thrive is to be efficient and innovative. Through innovation, the consumer electronics industry is moving us towards a new future, changing our relationships with our gadgets and with the world around us. And there are phenomenal amounts of money to be made along the way. A 2007 study by Gartner Inc. estimated sales of music through mobile devices at more than $13 billion during that year, and predicted a corresponding figure of $32 billion by 2010. The games market is very healthy, too: the world's gamers spent $35 billion on the games alone in 2007. And a game called Halo 3 generated $300 million of sales in its first week on sale during September that year – this is more than many Hollywood blockbusters take in their entire first run.

The £165 billion consumers spent on electronic gadgets in 2007 was £8 billion more than the equivalent figure for 2006. Spending on consumer electronics has risen year after year, although the rate of increase has been slowing. This deceleration seems to be due to falling prices, not reduced sales or waning

Much of the assembly and testing of electronic gadgets is done in China, in the hands of skilful labourers in huge, high-tech factories.

consumer enthusiasm, since unit sales are up. Doing particularly well in recent years are personal media players, smartphones, LCD and plasma televisions, DVD players and recorders, and hard-disk–based personal video recorders.

Ten years ago, the hottest sellers were CD players – and do you remember the MiniDisk? Video cassette recorders were still in vogue, but there were no hard-disk–based video recorders. Mobile phones were basic and still owned by the minority of the population, even in rich nations. There were no digital television services, no flat panel televisions, no mp3 players. The DVD player was just about bursting onto the scene, but as with most new, 'disruptive' technologies, the prices of the early machines were high. Prices of newly introduced technologies typically start high relative to the level they achieve if and when they achieve mass appeal. The people who buy into new technologies when they are still highly priced are called early adopters, whom Guardian Unlimited journalist Steven Poole has referred to as the 'tragic, unsung foot-soldiers of the technology revolution'.

I am not an early adopter. I can only afford to buy into a particular technology some way down the line. I have already confessed to being a tech junky, eager for continued progress to marvel at. But even I find it annoying that we have to keep forking out to upgrade to generation after generation of new technologies – and the content that goes with them. How many albums have you had to buy on CD that you already owned on vinyl? And how many of your treasured films on VHS have you bought again on DVD? And if you haven't already, will you buy some of those films yet again when you get hold of a high-definition DVD player? Let's face it: you don't want to have a VHS player and a DVD player and a high-definition Blu-ray disc player. While the Blu-ray player is 'backwards compatible' – it can play DVDs – will that necessarily be true of the next generation of devices to play our films? It's annoying to know that you have spent good money on dozens or hundreds of films that you will no longer be able to watch. You could choose to step off this unstoppable treadmill: photographs that exist only in albums, music collection on scratched vinyl, films on VHS cassettes. Does that idea fill you with nostalgia or make you pleased that things have progressed? How will you feel in another ten years? How many times will you have to upgrade before then? If anything, progress is speeding up; but it will only continue if we are willing to part with our money to keep fuelling it.

Making It

The ultimate beneficiaries of the money we spend on consumer electronics products are the companies that design and build the devices. The biggest companies are so successful that they have become household names around the world. Among the best known are Japanese companies such as Sony, JVC, Toshiba, Sharp and Hitachi; American companies such as Intel and IBM; South Korean companies such as Samsung and LG Electronics; and European companies such as Philips. Most of these organisations are truly multinational, with offices and manufacturing plants in far-flung parts of the world.

At the heart of any consumer electronics product lies the integrated circuit (IC). The processor, the RAM and ROM memory chips, and flash memory and the light-sensitive CCD sensor in a digital camera are all ICs. The importance of integrated circuits in consumer electronics means that their manufacture alone is a multi-billion dollar industry. There is a variety of approaches to the manufacture, or fabrication, of ICs.

Some companies, such as Intel, make integrated circuits but don't make finished consumer products. Intel is the world's biggest semiconductor manufacturer (by revenue): in 2007, the company brought in a total revenue of more than $38 billion (£20 billion), with net income of $7 billion. But there are no Intel-branded computers or mp3 players. Instead, there are hundreds of millions of devices with Intel's chips 'inside'. Other hugely successful manufacturers of integrated circuits include Infineon, Renesas and the Taiwan Semiconductor Company. These companies are not familiar to most consumers, because their products are normally hidden from view inside products sold under more familiar brand names. Some companies make ICs both for their own products and for others'. For example, Samsung chips are found in Samsung-branded televisions, mobile phones and media players, but also in certain models of Apple's iPods and in Microsoft's XBox 360. Some IC 'manufacturers' specialise in the design and marketing of integrated circuits, but don't have their own fabrication plants, so they outsource the production of the chips to specialist firms. These companies are described by the wonderful term 'fabless manufacturers'.

Of course, there are hundreds of components besides integrated circuits inside any high-tech electronics product, so even a company like Samsung, which makes ICs for its own brand products, outsources most of the components inside their finished products. Typically, the integrated circuits – along with the hard disks, wiring, circuit boards, display screens, batteries and keyboards – are assembled into finished products quickly and efficiently by companies under contract to the branded manufacturers. This approach is called electronic contract manufacturing. The world's biggest contract manufacturer is Taiwanese company Hon Hai, which trades under the name Foxconn. In 2007, Hon Hai had a revenue of more than $50 billion. In its massive factories, Hon Hai's workers assemble many familiar products, including Apple's iPhone, Sony's PlayStation 3 and Nintendo's Wii games console, as well as computers for Hewlett-Packard and Dell. Although the big contract manufacturers typically have dozens of assembly plants in several parts of the world, they are increasingly investing in China. Foxconn has most of its employees – nearly half a million – in China; nearly 300,000 of them work in a single factory. Assembly workers in China are known for their exceptional dexterity, high productivity, flexibility and adaptability – and above all their low labour costs.

As with any global industry, we the consumers are so far removed from the production lines as to be blissfully unaware of the conditions that some factory workers endure

Sometimes, the low costs of labour can come at a high price of a different kind. Our desire for low prices and the fierce competition within the industry put tremendous pressure on manufacturers. Sometimes that can lead to manufacturing costs being cut – courtesy of the workforce. As with any global industry, we the consumers are so far removed from the production lines as to be blissfully unaware of the conditions that some factory workers endure. The most high-profile allegations about sweatshop working conditions in the consumer electronics industry came in a report by the *Daily Mail* in 2006. The newspaper published allegations that workers at Hon Hai's factories in China were working 15 hours per day for $50 per month, and were living in large impersonal dormitories. Photographs of the dormitories and the military-style management became headline news – briefly. Some of the company's factories were allegedly situated in huge complexes with living quarters, sports facilities and shops – all surrounded by barbed wire fences. Workers were apparently allowed few personal possessions, and had only buckets in which to wash their clothes. The company denied the allegations, and stated that the conditions in its factories were above the average for China. I am not sure whether I should feel relieved or horrified by that. Among several companies whose products were being assembled in the factory were Apple; they launched an investigation. Apple found a few minor violations of its supplier code of conduct, but not enough to stop its relationship with Hon Hai.

New Wealth

China's association with the consumer electronics industry began in the 1980s, when several multinational companies opened factories in China to manufacture televisions, video cassette recorders and memory chips. Because of the skilled workers and low costs found there, more companies followed, and built hundreds of factories making cables, circuit boards and electronic components. Most electronics plants in China are located in the southeast and on the island of Taiwan (which is governed by China), so that they are close to Japan and Korea. Some of China's contract electronics manufacturers have taken to designing their own products, and they may begin to threaten the dominance of established companies in Japan, South Korea and the USA. Electronics represent a sizeable portion of China's overall wealth, which has been growing rapidly since the late 1970s. China is third or fourth in the world in terms of gross domestic product (GDP, total income for the nation). China's GDP has been increasing at around 10% per year. With such a huge population, covering an immense area, rapid change and growing prosperity in one part of China will take a long time to filter down to everyone. In terms of *per capita* GDP' (income per citizen), China still ranks around 80th in the world.

As a result of the new wealth in China, there are huge numbers of people with newly acquired disposable income and sophisticated tastes, more than willing to buy into consumer electronics devices, and China itself is becoming a market for the products that were previously only assembled there for people in other parts of the world. It is also becoming a major force in software engineering. With programmers' wages typically one-tenth of that paid to similar workers in the USA and Europe, China has a competitive edge. Brazil, Russia and India are in a similar economic phase, though presently not to quite the degree found in China – all having recently shifted towards capitalism, and all heavily involved in the consumer electronics industry in one way or another. All four countries also have important resources they can exploit, including oil. Economists often group these four countries together using the acronym BRIC. The emergence of the BRIC nations as major players in the global economy will help to buoy up the consumer electronics industry for years to come. Eventually, they will no longer be 'emerging markets': the economist who first used the acronym BRIC, Jim O'Neill at Goldman Sachs, has suggested that by 2050, these four countries will be the dominant economies on the planet.

Making Chips

A typical microprocessor for a PC retails at a few tens of pounds. A processor in a mobile phone or a media player is usually cheaper. And a standard integrated circuit used to convert an analogue sound input to a digital signal is much cheaper still. And yet these beating hearts of our gadgets are incredible feats of micro-engineering containing as many as half a billion components.

Like any electronic circuit, an integrated circuit is a collection of interconnected components – such as resistors, capacitors, diodes and, most importantly, transistors. In an integrated circuit, the transistors and other components are all integrated onto a single crystal of semiconductor – normally silicon. This is why microprocessors and other integrated circuits are sometimes called 'silicon chips'. There are other semiconductors that work equally well, but silicon is the cheapest and most abundant.

To understand what a semiconductor is, and why it is important, it is useful to contrast its behaviour with that of good and poor electrical conductors. The wires that connect the processor to the rest of a device are made of metal, because metals are very good conductors of electricity. However, you cannot turn off their conductivity – they are 'always on', which is no good if you want a two-state system capable of representing the binary digits '0' and '1'. Conversely, most non-metals are poor conductors: very little current can ever flow through them, so they would always be 'off' in a digital circuit.

Semiconductors are normally poor conductors, but they can conduct electricity very well if, for example, an electric field passes through them, or if they are heated or illuminated. The point is that the conductivity of a semiconductor can be controlled, normally so that they switch between 'very good' and 'very poor'. This is why these materials are ideal for binary digital electronics. Furthermore,

Silicon is a silver-grey metallic element, but unlike most metals, it is a semiconductor, not a conductor. Silicon used by the electronics industry is ultra-pure.

the conductivity of a semiconductor can fluctuate between these two states very rapidly. The tiny transistors on a modern integrated circuit can switch billions of times every second. Semiconductors are also the basis of small lasers found in CD and DVD players, image sensors in digital cameras and camcorders, and LEDs (light-emitting diodes) also common in many consumer electronics products. As we'll see, the semiconductor industry will remain a fundamental part of the consumer electronics industry for many years to come.

In its lifetime, the plant will **produce** tens of millions of **microprocessors, each with more** than **400 million** transistors

An integrated circuit is made from a square or rectangle of silicon called a die – because it is a small chunk of a larger piece that has been cut up, or 'diced'. That larger piece is a slim disc of silicon between 2.5 centimetres (1 inch) and 30 centimetres (12 inches) in diameter, and is called a wafer. The circular, silicon wafers are sliced from cylindrical ingots of highly purified silicon called boules. Imagine a thick sausage (the boule) being cut into hundreds of slices (the wafers) and then cut up into small pieces (the dice). Silicon wafers are normally about 5 millimetres (0.2 inch) thick, and each die is a small, flat rectangle (not a cube, the shape the word 'dice' normally conjures up). The length of a die is normally between 5 millimetres and 2 centimetres, and each die is identical to all the others cut from the same wafer.

From Silicon to Chip

The process by which a wafer of nearly pure silicon becomes a hundred or more integrated circuits is a complex, carefully controlled, high-tech affair that requires great precision. The whole procedure has more than three hundred distinct stages. Transistors and other components are built on the wafer's surface by adding and removing material, and by changing the electrical properties of tiny regions within the silicon. First, a machine lays down a thin layer of silicon dioxide on top of the wafer. On top of this, another machine deposits a metal, or sometimes a type of silicon with similar properties to a metal, and called polysilicon. The resulting 'metal-oxide-semiconductor' sandwich is made into electronic components by etching away the two new layers in just the right places using a laser. All of this is carried out in clean rooms – areas free of almost any dust, airborne microbes and chemical vapours that might compromise the purity of the silicon or the accuracy of the procedure.

The most important component on an integrated circuit is the transistor, and the most common type of transistor today is the MOSFET (metal-oxide-semiconductor field effect transistor).

In October 2007, Intel began producing integrated circuits at a new fabrication plant in Arizona, USA. In its lifetime, the plant will produce tens of millions of microprocessors, each with more than 400 million transistors. Two thousand of those transistors laid end-to-end would be needed to match the width of a human hair.

Within a MOSFET are specific regions called the source, the drain and the gate, each with particular electrical properties and with connections, via the metallic upper layer, to other components on the chip. Electric current flows between the source and the drain when the transistor is 'on' – but it can do so only when there is a voltage applied to the gate. There is a threshold voltage, below which the source-to-drain current will not flow, and above which it will. This is how a MOSFET can be a switch – a two-state system ideal for representing binary numbers and for carrying out binary mathematics.

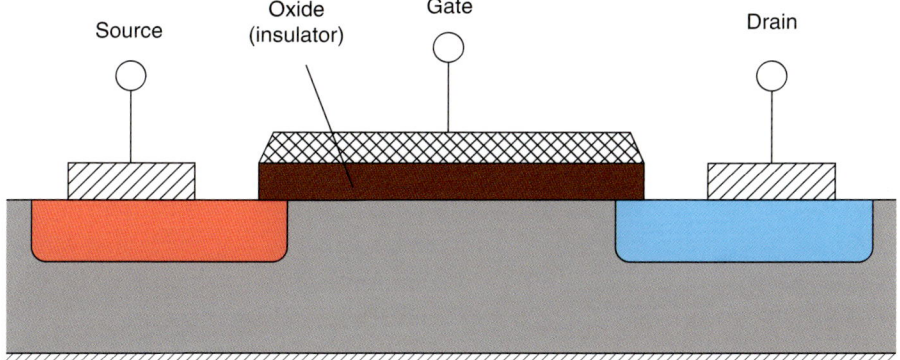

A shiny 30-centimetre wafer of silicon, onto which silicon oxide and metallic polysilicon have been layered, to make millions of transistors and other components in complex circuits. This wafer consists of several hundred individual central processing units (CPUs).

PHOTO COURTESY OF INTEL

Content, Services and Software

Profit from hardware often comes from the service providers or content providers. Hardware is often subsidised by the content or service provider – mobile phone handsets and video game consoles being two good examples. Manufacture is just the beginning of the road for our consumer electronics products. Companies that provide services – for example, retail outlets, Internet service providers, and those that build and maintain mobile phone networks – are an essential part of the industry. Without them, consumers would not be able to buy gadgets at all, or would not be able to use them effectively. Revenue in this sector is generated from licensing, advertising and subscriptions. The potential rewards are great here, too: Internet advertising revenue was estimated at more than $20 billion in 2007 . . . in the USA alone. It is clearly worth companies' while to invest this kind

Close-up of a silicon wafer, showing individual CPUs.

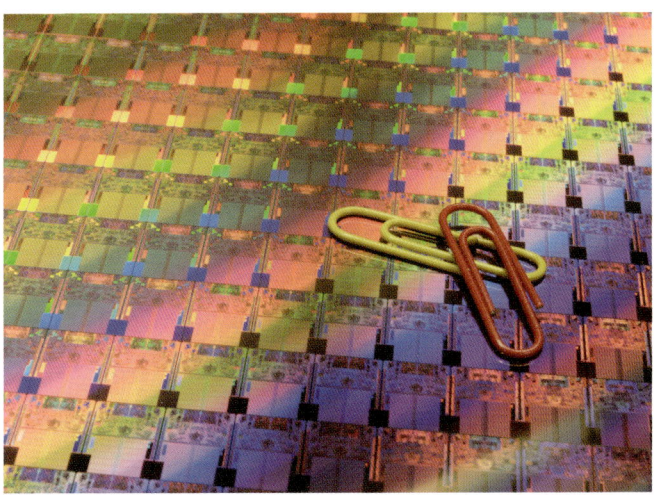

of money – so you could see the Internet as just another way to make us part with more of our money. However, if the opportunity to make money had never existed, then the Internet would still be limited to military and academic institutions. Just as the development of hardware depends upon end-users paying to upgrade, so does the development of the Internet infrastructure and services. If we want it, we have to pay.

One of the most important services available to consumers who own gadgets is the provision of connectivity. The rise of the connectedness of our gadgets – and in particular, access to the Internet – has provided fertile ground for many other types of service that can be accessed via personal computers or other connected devices. There are rich pickings for any business that can tap into the needs or desires of those of us who use mobile phones and frequent cyberspace. Mobile phone network providers are rich enough to bid billions for the licences necessary to provide their services, for example. And Google, which provides the engine used in more than 60% of the world's web searches, garnered advertising revenues of more than $12 billion in 2006. In that year, it bought online video portal YouTube for $1.2 billion. The company is so profitable that in the same year, it was easily able to survive losing 8.5%, equivalent to $13 billion of its total value, in a single day. The loss was in stock value, after investors became worried about the future of online advertising and at the same time, were concerned about Google's refusal to hand over search details of its users to a US federal court. In 2007, Google offered $3.7 billion for online advertising giant DoubleClick. After challenges to the legality of the purchase by other big players, the deal was finally approved in December of the same year. The Google empire, like the Microsoft empire, has amassed immense profits, but like Microsoft, it has ploughed huge amounts into philanthropic projects to combat poverty and climate change. Vulgar capitalism or a necessary force for change? I don't know.

The World Online

Our lives are becoming more and more dependent upon the Internet. There is email and online banking; owners of digital cameras can obtain prints by uploading their images; and manufacturers of consumer electronics devices provide documentation, additional software and firmware updates online. There are also companies offering safe online storage, as back-up of important documents, and to enable 'remote computing', with which consumers can produce and edit documents from any connected device. Online retail outlets and price comparison sites make buying consumer electronics much easier, too – and provide reviews and information about the available products. Then there are streaming radio and television stations and music download sites, social networking. While some of these services are provided by online communities, the majority are supported by advertising or by the revenue generated by the service they provide.

Global Network

When you make a call on a mobile phone, radio waves carrying digital signals pass between the phone and the nearest base station, typically a mobile phone mast. The wireless radio link between phone and mast is an example of a 'last mile' technology, also called 'last kilometre' technology.

Each mobile phone mast covers an area called a cell – this is why mobile phones are sometimes called cellphones. The masts are expensive to build and maintain, and the transmitters they are connected to use a good deal of power. They are connected to much larger telecommunications networks via two-way wireless or wired links that ultimately allow any phone, mobile or not, to connect to any other. The enormous costs of running these services, and updating them as new technologies become available, are ultimately borne by the end-user, via call charges or monthly network charges. Despite the substantial investment necessary in deploying last mile technologies, there is a tremendous economy of scale involved, because of the numbers of consumers using the available services. This is good for consumers, who pay reasonable rates for an important and valuable service, and good for mobile phone companies, which can make large sums of money despite keeping prices relatively low.

Just as the customers of a mobile phone company or Internet service provider pay for their last mile services, the phone companies and service providers are themselves customers of organisations that build and maintain the larger, nationwide and international high-capacity networks that underlie today's connected world.

The term 'last mile' is used in the same way to describe technologies that connect end-users to the incredibly fast, high-volume 'backbones' of the Internet. There are a variety of solutions available for delivering last mile Internet access. DSL (digital subscriber line) technology utilises high-frequency signals passing along existing telephone wires that originally carried only lower-frequency telephone signals. The underground cable network is another popular last mile solution. Cable Internet service providers build upon and extend existing cable television infrastructure to provide high-speed Internet access. DSL and cable Internet access are profitable enterprises – again, through economy of scale, users pay reasonable amounts, normally as monthly subscriptions. That is, they are reasonable when the service is consistent.

Optical fibre provides an excellent solution for the last mile connection, but massive investment is required in bringing it to every household and business premises

Increasingly, broadband companies are offering three services through the same 'pipe' – either cable, telephone cables or optical fibres. This 'triple play' model consists of digital on-demand television, broadband Internet access and telephone. The telephone calls are made using the 'voice over Internet protocol' (VOIP). Many Internet users will be familiar with VOIP as a way for making cheap or even free calls over the Internet. However, the quality of these calls can suffer during times of high Internet traffic or across a broadband connection of limited speed. Along dedicated last mile connections, however, the quality of VOIP telephone calls can be guaranteed, because the triple play company owns the last mile; connections beyond that – across the Internet at large – have enough capacity to guarantee the quality overall.

Optical fibres provide a very fast and high-capacity route for Internet, telephone and television traffic. They are used extensively in the Internet's backbone infrastructure. Increasingly, the same technology is being rolled out to homes and businesses, by service providers eager to break the near monopoly of last mile delivery that has been enjoyed by companies offering cable and DSL connections. Optical fibre provides an excellent solution for the last mile connection, but massive investment is required in bringing it to every household and business premises. There are several companies developing strategies to achieve this, but it is likely to be ten years before a majority of households gain Internet access through optical fibres. WIMAX is a last mile wireless technology, which can be accessed from roof-mounted antennas or in the open air from suitably equipped mobile devices. Wi-Fi – used for localised wireless networks – can be regarded as a 'last few metres' technology. It only gives access to a router already connected via DSL or cable services.

Something to Do

As well as the necessary services, content is also vital for many of our gadgets. What good would our media players, DVD players, digital televisions and Internet-ready PCs be without web pages, music, television programmes and films, for example? Revenue in this sector, too, comes from licensing, advertising and subscriptions, as well as direct sales. Not all content is good: the freedom of the Internet allows people to peddle pernicious content such as child pornography, for example.

Such is the importance of consumer electronics in our lives today that even the huge media corporations that have existed for decades are increasingly dependent upon it – although, of course, that dependency works both ways. It is no surprise, then, that companies that produce films, news and television programmes have merged or signed agreements with both manufacturers and service providers. Time Warner is the largest media conglomerate in the world, and among its subsidiaries are news portal CNN, Internet service provider AOL and cable television company Time Warner Cable. In 2003, mobile phone manufacturer Ericsson signed an agreement with RealNetworks, a company that delivers music and video over the Internet. The Sony Group is another example: the Sony corporation consistently appears in the top twenty consumer electronics manufacturers, while Sony Pictures Entertainment, Sony Computer Entertainment and Sony BMG provide film, video games and music respectively. Many people worry about the growth of media companies to a size where they have the power to influence governments as well as the minds and the culture of consumers worldwide.

Making It Work for You

Software is another vital part of the consumer electronics industry. Firmware, typically stored in ROM or flash chips and loaded at start-up, is part of the manufacturing process. Word processors and web browsers are services; games and encyclopaedias are more like content.

The operating system of a computer or mobile phone defies this simple classification. Although it is typically already installed when you buy a device, it is added after manufacture – and it provides services to the consumer, such as the ability to connect to the Internet and to store and access information on the device. It may also provide content. Some devices, such as digital cameras, have rather basic operating systems – only slightly more sophisticated than simple firmware. More powerful operating systems are found in mobile phones and personal computing devices. The two most widely used operating systems for personal computers are, of course, familiar to hundreds of millions of consumers: Microsoft's Windows and Apple's Mac OS. There are versions of these adapted for use in smartphones, ultraportable PCs and handheld computers. The iPhone, for example, runs a slim version of Mac OS, while Pocket PCs run Windows Mobile.

Most basic mobile phones run simple operating systems that are written by the phones' manufacturers. Smarter phones run more elaborate and powerful operating systems, such as Microsoft's Windows Mobile and Symbian OS – the latter being produced by a consortium of mobile phone manufacturers including Nokia and Sony Ericsson.

Windows, Mac OS and Symbian OS are 'proprietary' operating systems: they remain in the ownership of the companies that produce them. The software is licensed to the 'end-user', and there are limitations placed on what that end-user can do with them. They are normally ready installed in a computer or mobile phone, but that does not mean they are free. For example, for every Windows-based computer, the manufacturer – and ultimately, the consumer – pays Microsoft a licence fee. Mac OS similarly brings money in to Apple when it is ready installed in its hardware. And the consortium that developed Symbian OS receives fees for every mobile phone purchased that uses it. The popularity and wide compatibility of Windows – along with the increasing sales of personal computers – has led to market dominance for Microsoft and made it one of the richest corporations in the world.

Software packages with particular functions – such as Microsoft's Office productivity suite or Apple's Final Cut Pro film editing software – are called applications. These run 'on top of' the operating system. Some applications may be ready installed in a bought system, or they may be installed afterwards – but once again, most of them are money-spinning proprietary software. There are proprietary applications that are free, but they still come with restrictions on their use, and some are offered free of charge but still generate revenue. For example, Apple's popular iTunes music and video player is a free proprietary application, but it makes it possible for users to access the online iTunes music store, a very profitable venture. Still other free proprietary software is ad-based, or has hidden abilities to track users' web surfing behaviour, the resulting information being a valuable asset.

You might think that the producers of these proprietary software have us over a proverbial barrel: we need software, so we have to pay, right? Well, there is an alternative to the proprietary approach: free software typically produced by online communities of enthusiastic software engineers. Here, the word 'free' means much more here than 'without cost'. The programmers publish the software's source code – a line-by-line listing of a programs' commands – and anyone is free to peruse, edit, adapt and improve it. As a result, non-proprietary software is described as 'open source'. The most important open source operating system is Linux, a fully featured software platform that has begun to compete with Windows and Mac OS. Embedded Linux, a version of this operating system written for portable gadgets and dedicated devices such as network routers, is already widely used. Android is a Linux-based open source operating system and software tool kit for mobile devices, developed by a consortium of 34 companies including, notably, Google. For applications, too, there are open source alternatives. OpenOffice is a very popular and free productivity suite that is available for free download. The documents, spreadsheets and presentations produced by OpenOffice are compatible with all the major proprietary office software suites, including Microsoft Office. Similarly, the widely used open source image editing software GIMP does the same as the design industry's standard, but expensive, proprietary application, Adobe's Photoshop. Proprietary software comes with technical support from the company that sells it. Support for open source projects is found within the open source community – it is certainly available, but typically a little harder to find and understand for the non-technical user.

One of the features of open source projects is that they empower consumers – albeit those skilled in programming – to have an active role in the consumer electronics industry. End-user contributions are playing a major part in many other facets of the industry, most importantly in the production of online content. Blogging (producing 'web logs'), podcasting (producing Internet radio shows for download), social networks, wikis (collaborative websites) and online discussion forums all enable ordinary consumers to publish their own material easily to a

Software can be written in any of a number of human-readable programming languages. The result is called the source code, and it must be 'compiled' from a readable form into a binary form before a computer can carry out its instructions.

```perl
#!/usr/bin/perl -w

use strict;
use Gtk2 '-init';

# Create main window:
my $mw = Gtk2::Window->new('toplevel');

# Create main positioning table:
my $t = Gtk2::Table->new (1,2,0);
$mw->add($t);

# 'Hello world!' label:
my $label = Gtk2::Label->new('Hello world!');
$t->attach_defaults($label,0,1,0,1);
$label->show;

# Button to quit:
my $button = Gtk2::Button->new('Quit');
$t->attach_defaults($button,0,1,1,2);
$button->signal_connect( 'clicked' => sub { exit } );
$button->show;

# Main loop:
$mw->show_all;
Gtk2->main;
0;
```

global audience. As a result, many consumers are rapidly becoming producer-consumers, or 'prosumers' – a term coined by futurologist Alvin Toffler as long ago as 1980. The activities of Internet prosumers knit rather well into the existing, content-driven, link-powered World Wide Web. Rather than reducing revenue for the professional content providers, the kind of mass collaboration found at sites such as Wikipedia, and on social networking sites and blogs, only serves to integrate the online experience in people's lives, makes them feel involved in the digital age and encourages them to invest more time in their online presence. This in turn offers new opportunities for service providers, content providers and, of course, advertisers. Economist Eric von Hippel refers to this ethos of end-user participation as 'democratising innovation'.

Risky Business

Despite the vast amounts of money involved in all sectors of the consumer electronics industry, the profit margins for even the big players can sometimes be rather slim, and can even become spectacular losses. This may come as some surprise, since you might think that manufacturers could set prices as high as they please, given the desirable nature of the finished products and the eagerness of consumers to buy into the next big thing. But the costs involved in manufacturing integrated circuits, display screens and hard disks are huge. The panasonic corporation's new plasma display plant in Japan, opened in 2007, cost $2.4 billion to build; Intel's new microprocessor plant cost $3 billion. With such massive investment required in new developments, there is a great deal at stake.

Competition is not only about price. It is also about **making a company's products more desirable** than similar devices **offered by that company's rivals**

Competition is extremely important in this industry, and many companies have failed and folded. More often than not, competition drives down retail prices in order to attract customers. Companies have to take risks: huge manufacturing plants are only economical if output is high, and only profitable if sales are high. Competition is not only about price. It is also about making a company's products more desirable than similar devices offered by that company's rivals. As a result, features and performance are important and research and development is extremely well funded. Companies generally spend about 8% of their total revenue on it. The research and development activity of consumer electronics manufacturers is generally split between improving the performance or efficiency of existing technologies and developing new ways of doing things: innovative solutions to existing problems. Occasionally, two or more companies form alliances, pooling their spending on R&D for technologies that are particularly important and mutually beneficial. The Cell microprocessor, for example, found in Sony's PlayStation 3 and Toshiba's high-definition televisions, was developed jointly by Sony, Toshiba and IBM.

Companies that provide services take risks, too: spending huge amounts on building and maintaining last mile infrastructure is a gamble when there are other providers equally keen to attract business, and new technologies arising that can usurp or displace existing ones. The popularity of voice over Internet protocol (VOIP), for example, has had a dramatic effect on the revenue of telecommunications companies. VOIP enables people to make telephone calls via the Internet or through dedicated networks that use the same IP packet-switched approach. It is free or virtually free, and has forced traditional telecommunications companies to drop their prices.

Do You Copy?

There are risks for content providers, too: there are winners and losers and there are ups and downs. Companies that produce video games, for example, can spend millions of person-hours working on each new release; not all games are successes. And in 1999, Time Warner made a loss of $99 billion – at the time, this was the largest loss a company had ever made in a single year. Content providers can also lose significant sums of money through piracy – the illicit reproduction of content that is under copyright. Digital technology makes it easy to make exact copies of software, images, documents, and music and videos. The video games industry is at a particular risk of losing money to piracy, since the consoles are normally loss leaders: they are sold at a reduced price, and money is made when consumers buy games. There are a number of ways in which consumers can easily acquire illegally copied games cheaply or freely. Likewise, there are many programs that make it easy for people to copy CDs, to make mp3 files they can play on their computer or portable media players or share with millions of people over the Internet, via file sharing websites. The same applies to DVDs and even proprietary software.

Piracy is a big issue in the digital age.

Most companies that deliver digital content attempt to mitigate the risk of piracy in a number of ways, known collectively as digital rights management (DRM). Some DRM systems limit the use of digital content, while others attempt to prevent customers making illegal copies. The limitations enforced on a particular product are normally the result of negotiations and agreements between copyright holders – typically, media companies – and service providers and hardware manufacturers. One common approach is to include meta-data in the files that carry the content – something like a digital watermark, which cannot be copied or which can only be copied in certain circumstances. Manufacturers can then incorporate DRM technology into their hardware, so that devices will only access files with the appropriate meta-data. Many video game producers also incorporate copy prevention software in their games. Apple's iTunes online music store has its own form of DRM, called FairPlay. Like most DRM mechanisms, it tacitly monitors and controls the use of music files – in this case, any files purchased from the iTunes Store.

One criticism of DRM is that it can make certain combinations of hardware and content incompatible. For example, Microsoft created a DRM system called 'PlaysForSure'. Digital music files processed with this system will only play on approved devices, such as computers running Microsoft's Windows Media Player and certified portable music players. Bizarrely, Microsoft's own portable media player, the Zune, does not play music with the PlaysForSure system. Instead, it will play music copied from CDs by its accompanying software application, and it will play music bought from the Zune online music store. It will not play content bought from Apple's iTunes music store, or from any other music download service that incorporates DRM.

But DRM can be 'cracked' – disabled or bypassed – either by people who want to make money from pirated goods or by those who believe that DRM is a restrictive practice that is not in tune with the freedoms afforded by digital age, and who are resentful of the high prices charged by content providers. A cat-and-mouse

A **cat-and-mouse game plays out**, in which manufacturers and **content providers repeatedly** invent new forms of **DRM**, and **hackers attempt to break it**

game plays out, in which manufacturers and content providers repeatedly invent new forms of DRM, and hackers attempt to break it. In 2005, Sony BMG imposed a DRM mechanism on some of its CD titles. Controversially, the CDs automatically installed copy prevention software if they were played in a computer – without the customer knowing. Unfortunately for many users, and ultimately for Sony BMG, the software opened up vulnerability in computers, leaving them open to attack from computer worms and viruses. Many users sued Sony BMG in a class action; the company lost and was ordered to pay compensation. While the use of DRM is perhaps an understandable reaction to the potentially huge losses that media companies could make, the result feels quite at odds with a future in which digital information will pass effortlessly around, enhancing our lives. There is considerable consumer pressure on content providers – particularly in the music business – to provide DRM-free downloads. The idea is that if content is offered at reasonable prices, the vast majority of consumers would be happy to pay – it seems like an attractive strategy, given that DRM can be cracked. Several major companies have begun offering music DRM-free. It is an uneasy time for the large media companies, and it may take some years to settle down into a status quo.

Consumer electronics manufacturers are often accused of another practice that can be used to reduce the risk of making losses: built-in obsolescence. The practice of deliberately making products with short lifetimes first arose in the 1930s. As an economic strategy, it is not without risk: consumers left with a product that fails too soon after purchase may buy from a different company when they buy a replacement – if indeed they do. It is possible that manufacturers and software companies hold back the release of new developments, so that they can gain as much revenue as they can from the current generation of their products. This kind of built-in obsolescence is perhaps specific to high-tech industries, but it, too, is a risky strategy that makes little sense in an age where up-to-date information about product development is so easily available.

Obsolescence is an inevitable feature of many of today's consumer electronics products because the rate of progress in the underlying technologies quickly makes products outdated whether or not they fail physically. If a fault develops, it is nearly always cheaper to buy a replacement product than to have the existing one repaired. The replacement will almost always be more powerful or have more features. In many cases, repairs may not be possible, because integrated circuits are often designed for specific products, so that there are no replacement ICs available – and anyway, being soldered firmly to fixed circuit boards, they are not easy to replace. In fact, more often than not, products are replaced before they fail; such is the desirability and availability of the next big thing. Without people's willingness to buy into the next exciting development, the consumer electronics industry would be much smaller, and progress much slower. Whether intentional or not, obsolescence helps sustain progress by keeping turnover high. This is little solace to most of us – except perhaps in the long run, as we steam ahead towards that elusive technological utopia.

Costing the Earth

Whether obsolescence is built into consumer electronics devices as part of a clever strategy or simply a natural side effect in a fast-progressing industry, one of its consequences is huge consumption of energy and materials. Our ever

Whether built-in or just an inevitable prerequisite and consequence of progress, obsolescence is certainly very real.

increasing reliance on consumer electronics is making true consumers of us all – in the widest sense of the word. The quantities of energy and materials needed to manufacture, transport and power our gadgets are phenomenal.

In 1980, the use of electronic gadgets accounted for just 5% of the electrical energy used in homes in the UK. By 2007, that figure had risen to 15%. A 2007 study by the Energy Saving Trust estimates that it could rise to 45% by 2020. This would mean we will need the equivalent of fourteen power stations just to run our computers and other gadgets – in the UK alone. These figures are carefully worked out – not simply plucked from the air. Of course, 2020 is a long way away, and the study's estimate could turn out to be wildly inaccurate: the actual figure could be much lower, if the relevant measures are taken by manufacturers, governments and consumers. Alternatively, it could be much higher.

Many people assume that the new generation of **flat panel LCD** and **plasma screen televisions** require **much less power**, but actually the **opposite is true** in most cases

Sadly, the trend in recent years has been that our shiny new digital gadgets are more, not less, power hungry than their predecessors. The best example of this is televisions. The average cathode ray tube (CRT) television draws less than 100 watts of power when fully working. Many people assume that the new generation of flat panel LCD and plasma screen televisions require much less power, but actually the opposite is true in most cases. At a screen diameter of up to 30 inches, LCD and plasma screens use about the same power as a similar-sized CRT. But much larger screens have become more affordable, and very popular as replacements for CRT screens. Most 40-inch LCD or plasma screens require about 350 watts. Some models use less power than others, but most consumers do not concern themselves with finding out which ones they are. Many consumers are buying into the new television technologies because the new screens are attractive – not simply because their old CRT television has stopped working. When I bought our LCD television, I researched which companies were doing their best to reduce the use of materials and power in their factories and to produce products that use power efficiently. I have read several pieces of research that suggest these concerns are shared by only a small minority of television buyers. And certainly, when I have visited shops to choose the television model for me, it was not easy to find out the power rating of the televisions. This has to change.

A large proportion of televisions – and computers, set-top boxes, home network routers and personal video recorders – are left on standby when they are not being used. Again, I went for a television that has a standby power of less than 1 watt. Standby power can drain up to a third as much as when the device is in full use. It is all too easy to leave a gadget on standby – and it is very convenient. Some devices are designed to be left on overnight – this is particular to digital devices, which can receive automatic updates – such as next week's programme guide or a firmware upgrade. The report quoted above estimates that by 2020, about 1.4% of all electricity used in the home will be simply wasted by devices left in standby mode. Plug-in chargers used with mobile phones and other portable devices pose a similar problem. Several surveys have found that many people leave their chargers plugged in and switched on even when the associated gadget is in its owner's pocket or bag. Some of the more recent chargers have tackled this problem to a certain extent, switching to a low-power mode when not actively charging. These new chargers use less than 1 watt in that situation, compared to about 4 watts for their predecessors. But still, with a few billion mobile phones and other portable devices worldwide, even 1 watt per device of completely unnecessary power, for several hours each day, really adds up.

It is not only the consumption of energy after purchase that is of concern. Often overlooked is the energy used in the production, packaging and transportation of electronic consumer products. Every time we replace an outdated or defunct piece of technology with a new one, we are unwittingly using large amounts of 'embedded energy'. And, as I pointed out at the beginning of this chapter, unit sales are increasing. There is also a significant upward trend in the number of single-person households, and if this continues, the total number of televisions, personal video recorders, set-top boxes, games consoles and computers will rise steeply. This may be good for the industry, but it is not good for the planet.

Energy use is not the only concern: the hunger for more and better products gobbles up oil and other resources at an ever increasing rate. In 2002, a team at the United Nations University in Tokyo published the results of a very detailed study of the processes involved in semiconductor manufacturing. The researchers worked out that 1.2 kilograms of fossil fuels, 72 grams of other materials and around 32 litres of water were required to fabricate a single 2-gram computer memory chip. Furthermore, around 400 grams of fossil fuel would probably be burned to produce the quantity of electricity needed to make the chip work during its lifetime. In 2004, the same team published a similar study, using a similar approach, in which they estimated that the manufacture of a 24-kilogram desktop computer with a (CRT) monitor requires 240 kilograms of fossil fuels, 22 kilograms of other materials and 1,500 litres of water. Similar studies carried out on other consumer devices would doubtlessly yield similar results – this industry is materials- and energy-intensive. Manufacturers are well aware of this problem, and several large companies have initiatives in place to reduce their environmental impact. It should certainly be a major factor in design of future gadgets. Fortunately, there are several non-governmental organisations that act as industry watchdogs. And just as fortunately, the information they produce is freely available thanks to the Internet, for anyone who is interested to find it.

What happens to all of the devices we no longer need, either because they have stopped working or become outdated? Each year, the world produces around 50 million tonnes of electronic waste, or e-waste: products that are no longer working or no longer needed. There are many schemes that make previously owned computers and mobile phones available to those who cannot buy the latest technologies, but such schemes are generally not well used. And there are many types of gadget other than computers and mobile phones that are not normally included in such schemes. During the 1990s, some European countries legislated against the disposal of electronic waste in landfill sites – although many devices still find their way there. There has been no blanket ban in North America, however – although certain US states have some laws that aim to help. For example, across Canada and in some US states, there is a ban on dumping televisions and computers, and a tax is added to the price of these items to cover their recycling, which is statutory.

Although there are recycling sites that can deal with consumer electronics products, the process is expensive and only a small fraction of e-waste actually ends up there. Instead, much of it finds its way to recycling facilities in certain developing countries – in particular, India, China and Nigeria. Workers in these sites sift through and dismantle discarded items to salvage precious materials, including gold and copper. One site – in Guiyu, China – employs nearly a hundred thousand people. But their work is not desirable, as the gadgets they disassemble once were: the 'recycling' takes place in the open air or in open shacks. Workers use small fires to heat circuit boards in order to melt solder, and they are surrounded by huge piles of computer carcasses, electronic components, metal and shredded plastic.

Whether it ends up in landfill sites or is recycled by people in Asia or Africa, e-waste presents another problem: it contains many toxic elements and compounds. Some of these substances are hazardous to human health, while some are potentially damaging to the environment. Both of these problems are particularly pertinent in the primitive recycling facilities in Asia and Africa. Tests of soil and groundwater around the sites reveal that lead, barium and mercury have leached into the soil, and a study of children around Guiyu found that more than 80% have blood lead levels above accepted norms. Again, there are non-governmental organisations that publish lists of which manufacturers are the 'cleanest' in terms of their use of toxic substances.

One of thousands of Nigerians involved in repairing and reselling imported used electronic equipment. Unfortunately much of the imported electronic equipment cannot be repaired and is instead dumped.

Many of the hazardous substances found in e-waste are subject to restrictions on export and import via the Basel Convention, an international treaty initiated by the United Nations Environmental Programme that came into force in 1992. In principle, that treaty should make it very difficult for e-waste to be transported across international borders. However, it is difficult to police, and the USA – apparently responsible for most of the e-waste arriving in Asia and Africa – has signed but not ratified the treaty. Investigations by the Basel Action Network, Greenpeace and the Silicon Valley Toxics Association have brought this export trade to public attention. There are other agreements that have been developed in an attempt to deal with the growing problem of e-waste. One way of reducing the hazardous nature of e-waste is to restrict the use of dangerous chemicals in the first place. In Europe, the Restriction of Hazardous Substances (RoHS) Directive came into force in 2006. This imposes strict limits on the concentration of six substances – including lead, mercury and cadmium – in the manufacture of electronic and electrical products. A related piece of European legislation, which became law in 2003, is the Waste Electrical and Electronic Equipment (WEEE) Directive. It puts the responsibility for disposing or recycling of e-waste on the manufacturers.

Despite these initiatives, and a new willingness on the part of some manufacturers to bring 'greener' products to the market, massive consumption of energy and materials is set to remain a big problem in the consumer electronics industry, as it will continue to be in several other modern technology-related activities. However disturbing is the outlook for the environment, the industry has a great deal of momentum, and it keeps on producing what we, the consumer, crave.

Digital devices, and our relationship with them, will surely continue to undergo rapid and continuous change. There are clear trends, which will be explored in the next chapter. By examining these trends, it is possible to piece together the main developments in the consumer electronics industry in the near future.

3 The Long Slow Death of the Desktop

The personal computer is about to become the victim of its own success – not today, not next week, but gradually over the next few years. After the CD player, the desktop PC was the first digital gadget that most of us grew close to; and it has been a workhorse of our digital lives ever since. It does everything digital, and it does it well. But now that it has taught us how to be digital, we will soon be moving beyond it. By looking at hardware trends, along with the changing ways in which we use our gadgets, it is possible to chart the coming downfall of the whirring box under the desk, which holds our data captive.

The PC might live on – inside your television. After the success of their child-friendly notebook computer, the Eee PC, Taiwanese company Asus announced in January 2008 that they would launch a 42-inch LCD television with full PC functionality. While this might seem like a ray of hope for the doomed PC-centric model of computing, it is symptomatic of the fact that the PC is retreating from the desktop and finding its functionality in other devices. Asus' product is a television first and foremost, not a PC.

In our **digital** age, a **simple but unpredictable idea** can relatively quickly lead to a **sea change** in the **behaviour of hundreds** of millions of people

I am being slightly dramatic here. It is unlikely that the stand-alone PC will disappear altogether any more than books or vinyl records will. There will always be a place for it. But things are going to change. In tomorrow's world, our digital lives will be much more decentralised than they are today. We will be able to access our information when and from where we need to, using a range of network-enabled gadgets called 'thin clients'. The revolution has already begun, and a new breed of thin clients will take no prisoners.

Trends – Into Tomorrow

Making predictions about what will happen twenty or thirty years hence is an imperfect science within any field of human endeavour. But in consumer electronics, it is perhaps especially so, such is the dizzying rate of technological progress and the fast-moving and competitive nature of the industry. Furthermore, in our digital age, a simple but unpredictable idea can relatively quickly lead to a sea change in the behaviour of hundreds of millions of people. For example, Google's innovative approach to Internet searching quickly attracted a loyal following when it was introduced in 1998. By 2000, Google's search engine had become the most popular on the Web. And by 2006, the verb 'google' was added to the *Oxford English Dictionary*. If you want to check that fact, just google it.

Social networking is taking over the world – who doesn't know Facebook?

Similarly, sites with user-generated content, such as Wikipedia, blogs and podcasts, and social networking sites such as MySpace and Facebook, have very quickly become part of everyday life for a large proportion of Internet users. These dynamic, inclusive, interactive sites are part of a makeover of our online experience; the result is often dubbed Web 2.0, and it is quite a different entity from the static, one-way sites that were all the Web had to offer when large numbers of consumers first began connecting to the Internet in the mid-1990s. Millions of us access and contribute Web 2.0 content using a range of connected devices, including mobile phones, certain games consoles and of course, our PCs. There are hundreds of websites that offer unsigned, amateur musicians the chance to sell their music online and reach huge potential audiences. These kinds of services have become significant features on the landscape of consumer electronics, and a regular and important part of many people's lives – but not so very long ago, they did not exist.

The Web 2.0

If you use the Internet, you cannot have failed to notice the rise of sites with user-generated content and with customisable elements. If you have a Google account, you can customise your homepage to include any of several hundred gadgets that give you live updates on news, travel and weather, tools that help you check for and write emails, browse your photographs stored at online sites such as Flickr or on your PC, and play games. There are dozens of similar sites with this kind of modular, customisable approach.

The Web 2.0 is not a 'new version' of the Web; it is simply a trend towards more dynamic websites and the emergence of much more user-generated content. It is characterised by collaboration between Internet users, and by social networking services that are hosted online, such as the hugely successful MySpace and Facebook. The emphasis is on customisation and on the fact that it is as easy for people to upload information to these sites as it is to download. These sites do arouse concerns about privacy. For example, some employers have begun routinely checking through the pages of job candidates on social networking sites. So, much against the spirit of the creativity and freedom of expression on the Web, what you write could affect your future prospects.

The Web 2.0 is very different from the static landscape of the Web just a few years ago, and is very welcome as far as I am concerned. The change is largely down to new approaches to programming applications that run on web servers. Most important is a programming language called XML (extensible mark-up language), which allows for dynamic content to be dropped into web pages easily. XML is also the basis of RSS (really simple syndication), which offers web surfers the chance to keep up to date with their favourite websites. An RSS 'feed' is a short document that automatically lists new documents on a large website, such as a news service. An RSS reader – typically integrated into a web browser – provides a quick and easy way to check for updated content.

While the very rapid and punctuated evolution of the consumer electronics industry makes prediction difficult over a long period, the picture is slightly different over shorter timescales – notwithstanding the arrival of the next Google- or Wikipedia-type phenomenon. It normally takes between three and five years for new manufacturing techniques to reach production level or for industry alliances to be forged. Once a product is finally on the market, it sometimes takes a year or two to make it past the stage when it is only bought by the keenest early adopters.

And, as pointed out in Chapter 1, the technology underlying consumer electronics changes much more slowly – so at least we have some idea what devices might be able to do, just not what they might look like.

In addition, some radical new ideas or devices can wait in the wings for years, well known and keenly watched by those who follow technological progress – 'bubbling under' until processing power or sufficient network infrastructure make them viable. In this case, prediction is just a case of guessing when the bubble might burst to the surface. For example, the first mp3 players appeared around the turn of the century, and quickly took the world by storm. But the mp3 format was designed as long ago as the late 1980s, and a vision of solid-state (no moving parts) digital music players had been bandied around even before that. Once the first mp3 players became available, prices remained high for a few years and it took a while before a trickle of early adopters became the flood of mainstream sales. So although we are living through a rapid revolution, it is not quite as rapid as it seems. Even once mp3 players became mainstream, changes in the technology were limited mainly to increased storage capacity, better device interfaces and the introduction of hard disk-based players – not that difficult to predict.

Keeping up to date with such technologies is still one of the **best sources** of predictions about the future of consumer **electronics**. The **other is trend-watching**

Of course, despite their knowledge of the coming of mp3 players, no one in the 1990s would have predicted the enormous success of Apple's iTunes online store – and, of course, equally successful online file sharing software – both legal and not-so-legal. Without them, the mp3 format's journey would probably have taken a different course and progressed at a different speed. Once again, it was the thinking 'outside the box' – about how we can use the technology, and not the hardware specifications themselves – that played the most important part. So the eventual success or failure of ideas that are in development is difficult to predict with confidence, but keeping up to date with such technologies is still one of the best sources of predictions about the future of consumer electronics. The other is trend-watching.

Computer scientist Edsger Dijkstra once wrote that when we build sandcastles, we can ignore the waves but we should watch the tide. In order to make informed predictions of what consumer electronics devices might be like and how we might use them, futurologists concentrate on trends, not so much on individual products. Perhaps most important are trends in hardware development – for example, improvements in processing speed or memory capacity; the capabilities and potential of any future device depend upon them. Then there is design – not only in terms of how devices look, but also how we interact with them: the interface. Finally, there are trends in how we use devices, and for what – as we have seen, this is the most elusive piece of the puzzle. However, put all of this together, and you can begin to build a picture of what lies ahead for our relationship with consumer electronics.

Of course, making predictions based on trends is guessing what will happen in the future based on what has happened in the past. In the same way, most professional gamblers betting on a horse race study the horses' form before placing a bet – they might still lose their money, but it's the best they can do. So I'm going to look at digital technology's recent form in the global marketplace and gaze into the near future. I'll start with hardware; here, the most important trends are increasing processing power, shrinking sizes, improving networking speeds and coverage, and falling prices.

Faster and Cheaper

The average computing power of personal computer CPU chips has increased exponentially ever since the invention of the microprocessor in the 1970s. The first real microprocessor was the Intel 4004, released in 1971. It had a 'clock speed' of 740,000 cycles per second (740 kHz). This number gives a direct indication of the chip's processing power, since instructions and mathematical operations are carried out in stages in the chip's circuitry, and one or more stages happen at each 'tick' of the processor's all-synchronising clock. The ticks are produced by a rapidly oscillating crystal. The clock speed of the processor in the first IBM PC (1981) was 4.71 million oscillations per second (4.71 MHz). This is an amazing six times the speed of the Intel 4004 – an increase of 500% in just ten years. By the mid-1990s, average speeds had reached 100 MHz – a twenty-fold increase in a little over ten years. And the year 2000 saw the first consumer-level chip to run at 1,000 MHz – 1 GHz, or one billion ticks per second. It took only two more years for personal computer CPUs to achieve 2 GHz and only another year after that to reach 3 GHz. Is there any other field of human achievement where you can see growth like this?

Clock speeds have not increased as steeply since 2003 as they did before then. But the number of instructions a chip can carry out per second (IPS) has continued to rise in higher-end consumer PCs. And this is a better indicator of a chip's raw processing power than clock speed. Generally of course, the higher the clock speed, the more instructions a chip can perform each second. But the exact relationship between clock speed and IPS depends upon a chip's circuit architecture. It was the introduction of dual- and multi-core processors that enabled processing power to continue increasing while clock speeds remained much the same. On a multi-core processor, two or more processors are present on the same chip, synchronised to the same ticks of a shared clock, and sharing computational tasks between them.

Processor Speeds

The Intel 4004 could carry out 92,000 instructions per second (92 kIPS), while a typical PC chip in the mid-1990s could carry out 100 million (100 MIPS). Intel's Core 2 X6800 chip (2006) – which consists of two processors on the same die running at a clock speed of 2.9 GHz – is capable of more than 27,000 MIPS. In 2007, Intel demonstrated a chip that could perform one million MIPS (a trillion calculations per second). The first computer to achieve that computing power was a supercomputer with 10,000 processors working in parallel. Intel's fingernail-sized chip had 80 separate cores running in parallel on a single die. Massively multi-core chips will be a very important factor in the future of consumer electronics, but their use will necessitate a real change in the way software is written. If applications cannot take advantage of the 'parallelism' of multi-core chips, then these chips provide no advantage.

At the same time as processing power is rising, chips are becoming cheaper to manufacture and cheaper to buy. In particular, with each increase in speed, the prices of previously high-end chips fall away dramatically, making them accessible to many more consumers. As a result, you can buy a cheap computer in 2008 that is much more powerful and also cheaper than top-of-the-range machines from just five years ago.

Similarly, a few years ago, laptops were luxury items – considerably more expensive than desktop computers. In 2007, not long after the 'hundred dollar laptop' made its début in developing countries, other small, low-cost but fully featured laptops came onto the market. The Asus Eee PC, which started at $200

How did we ever cope before processors could carry out a billion instructions per second?

Date	Processor	Instructions per second
1971	Intel 4004	92 thousand
1979	Motorola 68000	1 million
1987	Motorola 68030	11 million
1992	Intel 486DX	54 million
1997	Power PC G3	525 million
1999	Intel Pentium III	1.35 billion
2005	AMD Athlon FX-57	12 billion
2008	Intel Core 2 Extreme QX9770	59.5 billion

in the USA (and a different model for around £200 in the UK), and the Intel-supported ClassmatePC at around £110 ($200) were both aimed at children and low-end users. But both machines were fully featured modern PCs. Computer manufacturers Dell and Lenovo have also produced very cheap laptops with the developing world in mind. The processors in these super-value PCs are not as powerful as those in the mainstream models that sell for hundreds of pounds more, and they represent the biggest saving in the manufacturing costs. However, with clock speeds at between 400 MHz and 1 GHz, the processors are as powerful as those used in high-end consumer models just a few years ago.

The trend towards very low-cost portable PCs will almost certainly continue. And, as smartphones and other handheld devices become more sophisticated and more powerful, the two classes of device have begun to merge somewhere in the middle. Fairly soon, cheap, portable computers of significant processing power will probably be the norm. Instead of a small number of well-known brand names making relatively expensive computers and smartphones, the market will open up to a host of smaller manufacturers, driving prices down until the portable PC is a 'commodity'. This commoditisation is what happened to televisions, to DVD players and video cassette recorders before them, as well as to low-cost mp3 players. While commoditisation is often good for the consumer – because it drives prices competitively low – it can put pressure on large consumer electronics companies: those that spent billions of dollars on research and development to produce the earlier versions. As a result, it can even stifle progress.

Small and Cheap

The falling prices of computer components – including CPU chips – have led to the arrival of very cheap, fully functional portable PCs. This trend really began with an idea by a non-profit making organisation called the One Laptop per Child Association. A consortium of people and companies involved in consumer electronics got together and designed what became known as the 'hundred dollar laptop' – a rugged, networked, portable PC with a long battery life. It features a dual-mode screen, which can display at high resolution, in colour, with a backlit LCD – or, to save the battery, at low resolution in black and white by reflected light. It even has a wind-up handle to charge the battery.

The laptop was designed to be so cheap so that governments of poorer countries could buy them in bulk, ultimately giving one to each child. It has solid-state flash memory instead of a hard disk, and the whole thing was first available in 2007 for slightly more than $100. Governments in several countries quickly placed millions of orders. Interestingly, India rejected the idea, proposing instead an even cheaper version they could produce themselves – which should be ready by 2009. The point is that prices have fallen so much that this sort of thing is possible. And whether for profit or not, it seems likely that low-cost laptops will find their way into the developing world, encouraging computer literacy and the development of network infrastructure. Small devices like the hundred dollar laptop could make a big difference.

The 'hundred dollar laptop', may seem expensive in a few short years.

Increasing computing power and falling costs are the main driving force behind the digital revolution. It enables the manipulation of high-definition digital video, produces realistic graphics in video games and allows software applications to run more smoothly. So why is computing power increasing?

Shrinking Transistors

The growth in CPU processing power is due to a steady increase in the number of transistors that manufacturers and chip designers can cram onto a single semiconductor die. Not only is it the number of transistors present, but their proximity: when transistors are closer together, it takes less time for current to flow between them – it is (almost) as simple as that. The rise in the number of

Gordon Moore noted the trend in the number of transistors per chip in 1965. In 1968, he went on to co-found Intel – the most successful manufacturer of microprocessors to this day.

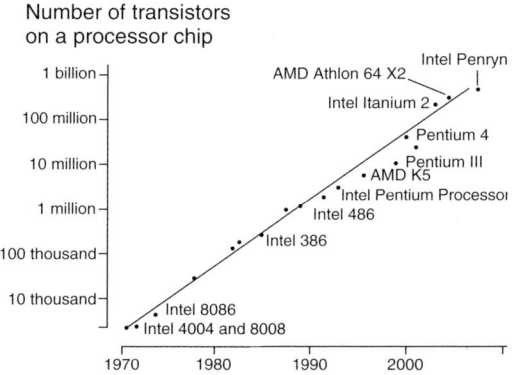

Number of transistors on a processor chip

transistors possible on a single chip has been meteoric. In fact, it has been an exponential increase.

In an article he wrote in 1965, electronics engineer Gordon Moore noted that the number of transistors that could be incorporated cheaply onto an integrated circuit had more-or-less doubled every year since the invention of the integrated circuit (in 1958). In that article, Moore announced that he could see no reason why the trend should not continue. He predicted that by 1975, the number of transistors on a single chip would reach a staggering 65,000 – up from around a hundred in 1965. This must have seemed like an enormous number at the time, but the introduction of very large scale integration (VLSI) around 1970 enabled the trend to continue, and Moore was proved right – more or less. He had to adjust his observation – what became known as 'Moore's Law' – in 1975, stating instead that the number of transistors doubles every two years, not every year.

VLSI is the process by which tiny components are made on the surface of a silicon wafer. The technology has continued improving, keeping up with Moore's prediction until now – and it shows no sign of slowing down just yet. In fact, most industry analysts – including Moore himself – consider that what was originally an observation has become somewhat of a self-fulfilling prophecy. Chip manufacturers tend to use it as a benchmark, and it is no coincidence that as soon as miniaturisation seems to be at risk of stalling with a particular approach, research and development teams work their hardest to come up with something to make it continue.

Cheap as Chips

Integrating more transistors onto the same chip also means that each transistor costs much less. In 1965, when Gordon Moore made his famous observation, the cost of each transistor on an integrated circuit worked out at more than a dollar. Today's chips are very densely packed with components and manufactured in greater volumes, meaning that the price per transistor has dropped to around one millionth of a dollar. Increasing the diameter of the silicon wafer from which chips are made has also played a role in reducing the price of integrated circuits, since with a larger wafer, more chips can be made at the same time.

Improvements in VLSI have ramifications throughout consumer electronics, not just in transistor 'density' and price of a CPU chip. For example, the light-sensitive chip at the heart of a digital camera is also made by VLSI, so the dramatic increase in the resolution of digital cameras – and the dramatic reduction in their price – also follows Moore's Law. It is a similar story for solid-state memory chips, since these are also integrated circuits made in the same way. The capacity of working memory (RAM) in bottom-end PC systems has risen steeply over the past twenty years and prices have fallen. Increased RAM is almost as important as higher processor speeds in determining overall computing power. The other important class of solid-state storage is flash memory. USB 'memory sticks' for PCs are flash-based, as are removable storage cards for digital cameras and mobile devices. The maximum capacity of flash chips has increased from 32 MB in 2000 to 64 GB in 2007. This is a 2,000-fold increase in just seven years. As flash chip capacities continue to rise, they have begun replacing hard disk drives as the main long-term storage device in PCs – a trend that is sure to continue. Flash-based storage is rugged and, unlike hard disks, has no moving parts, and so makes a good storage option. In June 2007, South Korean company Mtron announced that it had produced a flash storage chip that could read and write at up to 100 MB per second – faster than most hard disks.

However, hard disk capacity continues to rise, too, and prices continue to fall dramatically. Hard disk drives have gone from being measured in megabytes to gigabytes and then hundreds of gigabytes incredibly rapidly. In fact, the rate of increase in information density on hard disks is slightly faster than the rate of

Flash memory chips like these are common in digital cameras and media players. The 4 GB chip can store nearly six times as much information as the CD beneath it.

Moore's Law. The highest capacity disks of 2007 had an 'areal density' (binary digits per square inch) that was 50 million times as much as the first disk drive, made in 1956. Individual binary digits on a modern hard disk take up such a small space that researchers have to exploit phenomena of quantum mechanics, the strange physics that governs the atomic scale, to keep improving.

The biggest single improvement in hard disk design came in 1996, with IBM's introduction of giant magnetoresistance (GMR) into the design of read/write heads. In this quantum mechanical effect, a magnetic field changes the electrical resistance of a material over a scale of a few nanometres. The effect is also used in certain RAM chips. The scientists who discovered the GMR effect, French physicist Albert Fert and German physicist Peter Grünberg, won the 2007 Nobel

Prize for physics in recognition of their work. In 2005, hard disk manufacturer Seagate introduced hard disks that use another quantum mechanical effect, called tunnelling. Their TMR (tunnelling magnetoresistance) hard disk drives hold 400 GB on three platters (individual discs). Larger, terabyte (1,000 GB) hard disk drives have been available to consumers for a few years now, but these have always had many more platters. The latest development in hard disk design as of 2008 is 'perpendicular magnetic recording', in which bits are stacked vertically 'into' the surface of the disk instead of horizontally, one bit deep. Toshiba is one of the companies spearheading this new technology; in 2007, the company announced that by 2010, it will be producing portable computers with 1.2 terabyte hard disks (1,200 GB). The falling costs and increased densities of hard disk drives have made it possible for cheap, high-capacity hard disks to make it into personal video recorders, digital camcorders and personal media players.

The effect of all this growth of storage capacity at lower cost is the same as building ever wider motorways: it attracts more traffic, and will certainly become filled. Right now, I am not sure that I could fill 1.2 terabytes of hard disk space. But as capacities grow, and processor speeds, Internet speeds and camera resolutions increase, we will certainly be creating and storing even greater amounts of digital information in years to come. A feature film in high-definition can take up tens of gigabytes. So, loads of photos, music and high-definition films? Personally, I find it all a bit daunting – when I think of it all on my PC, anyway. More stuff than I can ever find time to enjoy, all held captive inside my desktop computer. It's not practical.

And so, despite the unbridled increase in hard disk capacity and the reduction in cost, it is not easy to predict the future of the hard disk in consumer devices. It is becoming possible to store a lifetime's worth of digital image, music and video files – along with more emails, e-books and documents than you could ever access – all on a physically small disk built into your PC or in a stand-alone hard disk back-up device. But there is also a dramatic growth in online storage options:

Designing the hard disk drive out of a consumer electronics product – or at least reducing its size – makes it slimmer, cheaper and less power-hungry

increasingly, people are opting to back up all their content remotely and automatically on hard disks at their Internet service providers or specialised online storage providers. That way, if the user's computer fails, or is stolen, the information is always available – and there is the added convenience of being able to access your information from any of a number of devices, from anywhere in the world. Currently, many Internet service providers offer large volumes of online back-up free. This is a good incentive not to switch providers: moving all your backed-up information across to another provider is a hassle at best. I never take that option for that very reason, choosing instead free or low-cost services from other companies.

The rise of online storage makes it possible for consumer electronics manufacturers to make products with very little built-in storage of their own that are nonetheless still fully featured. Designing the hard disk drive out of a consumer electronics product – or at least reducing its size – makes it slimmer, cheaper and less power-hungry. This approach is called 'thin client' computing. It has long been discussed, but it is only in the past few years that increased broadband uptake has made it a realistic option. The most significant example of thin client computing at the time of writing is the Zonbu PC, released in 2007. The main unit is a fully featured desktop PC sold (without monitor, keyboard or mouse) for just $99 (about £50). It has just 4 GB of storage built-in – as a flash memory chip – compared to typically 250 GB for even low-end PCs. Any files created on the PC – images loaded from a digital camera, or music files downloaded from the Internet – are automatically backed up to online storage, which is paid for by monthly subscription. With no hard disk and an energy-efficient CPU that does not require a cooling fan, it is small and completely silent. Over the lifetime of the device, the monthly subscription would probably add the same or more money to the price as the difference in cost between the Zonbu and similar machines with their own storage built-in. The prices of hardware and online storage will fall still further – and Internet access will become faster and cheaper.

One device that makes good use of online storage is Vudu – a set-top box with a difference. Vudu allows its owner to watch any of five thousand DVD-quality or high-definition films that are streamed directly via the Internet. The product comes with a hard disk that stores the first minute or so of each film, so that there is no delay in the beginning of a presentation, while the device connects to an online storage device that holds the rest of the film. The transition between the locally stored beginning and the streamed version of the film is seamless. Apple makes a similar device, called Apple TV. You can rent films – even in high definition – and when your rental period is over, the film is automatically deleted. This kind of approach may eventually do away with the need to buy DVDs. I like this idea: I have a hard disk filled with stuff I will never get around to watching, and a cupboard full of DVDs I have watched only once or twice.

So if people begin to find their hard disks becoming redundant, as they store more and more online, they may well be happy to do away with them. But another phenomenon – called peer-to-peer networking – might make it necessary to have large hard disks in PCs after all. Peer-to-peer (P2P) networking shares the burden of storing and delivering information among millions of users. So, instead of, say, a popular online video being delivered from a hard disk on a single network server to many Internet users at the same time, it can be accessed from the hard disks of any number of users. This *ad hoc* and distributed approach can reduce severe bottlenecks that occur when large numbers of people try to access particular

Looks like any other set-top box, but the Vudu can stream high-definition films across the Internet.

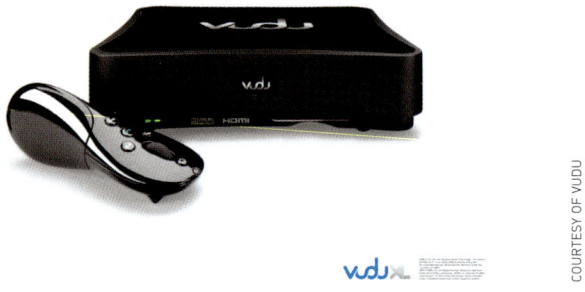

vudu XL

content at the same time. With the number of Internet users growing daily, and the wealth of online services expanding, these bottlenecks could become more common – and one day, the whole Internet might have to be based on some kind of P2P networking.

Whatever happens, hard disks will no doubt continue to play a major role in our lives for the foreseeable future. Most homes will probably have a central server – with a hard disk that will hold all a family's digital files – allowing the use of any number of thin clients anywhere around the home. This revolution has already begun, with 'network-attached storage' – familiar to anyone who works for a company as 'servers' or 'network drives'. In the past few years, networked storage devices have dropped in price and become available for home users. For as long as the PC remains the centre of our digital world, there may be some confusion as to why we might need a separate networked storage device in the home: why not just network the PC, which has a large hard disk anyway? However, networked storage in the home offers a chance to move away from the PC-centric model: instead of having a single home desktop computer and perhaps a laptop, cheap thin client computers can become integrated into every room, even built into walls and furniture. These devices will all be able to access the files of anyone who lives in the household, but will also connect with web-based content and personal online storage. This kind of distributed approach has been talked about for many years, but now, with networking technologies maturing and costs falling, it really could become a reality – slowly at first, perhaps.

Comparison of peer-to-peer (left) and server-centred networking (right). You can see how peer-to-peer networking can reduce congestion.

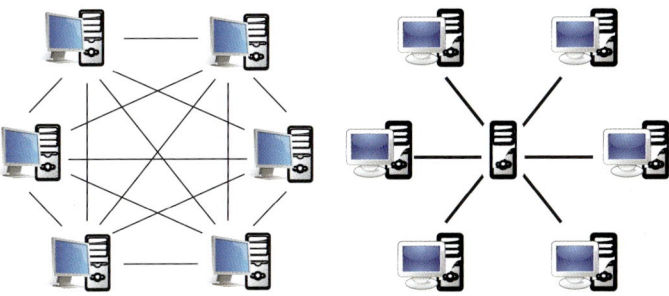

Peer-to-Peer Networking

In traditional networking, a computer that holds available content, such as web pages, blogs and videos, is called a server; a computer that accesses the content and services is called a client. Whenever a large number of clients attempt to connect to a server simultaneously, the whole network slows down. On the Internet, this can mean that the most popular sites are the ones most at risk of failing to deliver content consistently. With more and more of us relying on the Internet, and with more high-volume media services going 'online', the risks can only grow.

In peer-to-peer (P2P) networking, the notion of server and clients is replaced by the idea of many 'peer' computers, any of which might have copies of the required content. Any user looking to download, say, a popular video file, could receive the file in pieces, each piece coming from a different peer. Once assembled on the user's computer, the file would stay there, and could be accessed by someone else who wants that file at a later time.

Some P2P networks do also have central servers, but these only hold lists of which peer computers hold what files. One of the major advantages of P2P networking is that as the number of users grows, the reliability of the network actually increases – the opposite of what happens in the client-server model. P2P is most effective when delivering hefty files, such as videos, songs and large programs. Voice over Internet protocol (VOIP) and live video chats are also delivered via P2P networking.

Network-Attached Storage

Network-attached storage (NAS) is already becoming popular in the home where there are several computers – and there are plenty of models that, sensibly, have wireless capabilities. This idea is future-proof: it doesn't matter how many times you upgrade your computer or handheld devices, your information is always in one place. Music and video, photos and documents can all be accessed from any computer in the home network – including Wi-Fi-enabled portable ones. At the moment, NAS feels like an add-on to a PC-based approach. But it might be part of the downfall of the PC. It takes away from the PC the burden of storing large amounts of information, and makes that information readily available, even when your desktop PC is switched off.

Future portable devices will be thin clients, too, holding very little information on-board. Imagine having a media player that holds no music at all, but which can access your own files from anywhere. When gaining access to the Internet is as widespread and easy as making a mobile phone call – actually, more so – there is no reason why this cannot be. However, for the foreseeable future at least, portable devices also need significant back-up storage of their own, for use in places where network connectivity is poor. A portable media player that has no media stored on it, and which cannot access any media online is of no use at all. As flash memory continues to grow in capacity and drop in price, that will be the natural choice, as it is smaller and quieter than a hard disk – although hard disks will remain ahead in terms of volume and price for some time to come. The coexistence of all these technologies will further enhance one of the most important features of the consumer electronics: consumer choice.

Imagine having a **media player that holds no music** at all, but which can access your own files from **anywhere**

The emergence of online storage and thin clients – that will signal the death of the desktop PC – leads us to another major trend that is a driver of change in the consumer electronics industry, besides advances in processing power and storage capacity: improvements in networking.

New Connections

In the near future, our gadgets will be better connected to each other – within the home, in our cars and even in our clothes – and they will be better connected to the rest of the world, too, via faster and more dependable Internet connections. The number of networked devices is set to increase astronomically. The Internet is set to be so pervasive, and such an important part of every aspect of our lives, that we will always be connected – wherever we are, and in any of a large number of possible technologies. Within ten years, the Internet will be so integrated into our lives – and will extend into every part of our homes – that we will probably no longer speak of 'going online'. Technology journalist Walt Mossberg has compared the Internet of the future to the electrical supply system: we never say we are 'going on the electrical grid' when we plug in a toaster or use a hair dryer.

The most rapidly growing aspect of our 'connectedness' is wireless networking, whose recent popularity will almost certainly continue and grow. Many homes that have broadband Internet connections also have a wireless router, which extends Internet access to any wireless-enabled devices in the home. These 'Wi-Fi' networks are often difficult for non-technical users to configure and repair, and can be prone to interference and signal strength drop-off inside the home. But the quality of Wi-Fi networks is improving all the time, and prices are falling; there is no sign of them becoming anything but more numerous and more dependable. Wi-Fi connections will play a major role in our increasingly connected homes.

Already, many consumers view PC content on their televisions or listen to music on their audio systems via Wi-Fi links. Many people have wireless printer/scanners and wireless-enabled handheld computers, media players or smartphones.

Wi-Fi

Wi-Fi is the common name of the 'language' of wireless local area networks (WLANs). It is based on standards designed by the Institute of Electrical and Electronics Engineers – called IEEE 802.11. These were introduced in 1997, and allowed for transmission speeds of 2 megabits per second (2 Mbit/s) and a maximum range of 100 metres. The speeds and/or the range of wireless networks have been improving ever since. Many consumers will be familiar with 802.11b (1999) and 802.11g (2003); in 2009, the 802.11n standard should be implemented, allowing speeds of up to 248 Mbit/s and a range of 250 metres. There are many other variations of IEEE 802.11 in development, which will provide wireless Internet access from moving vehicles, or Internet telephony (VOIP), or wider Internet access in town centres.

The transmission speed, or throughput, of wireless networks depends upon the frequency – the number of oscillations per second – of the radio waves that carry the digital information. A radio wave oscillating more rapidly can carry more binary digits per second, and therefore provide a greater transfer rate. Wi-Fi uses 2.4 or 5 gigahertz (GHz) – although additional 'space' may be available to it in years to come, for example around 3 GHz. A different set of standards provides wireless networking at much higher frequencies: between 10 and 66 GHz. It is officially known as Wireless Metropolitan Area Network (WMAN), but it is more commonly referred to as WIMAX (Worldwide Interoperability of Microwave Access). It has greater range and transfer speeds than Wi-Fi, but the two standard sets will coexist, since they are designed for different purposes – although both will offer very cheap or free VOIP that will threaten traditional telephone companies and even mobile phone operators.

The best-known voice over Internet service is Skype. Normally accessed from a PC-based application, Skype allows any of its nearly 300 million registered users to hold telephone conversations and video chats via the Internet free of charge. You can make 'SkypeOut' calls to normal landlines and mobiles, too, at a cost. Recently, Skype has been available as an application that will run on handheld devices that have Wi-Fi connectivity. And in November 2007, the Skypephone appeared in several countries.

However, the mobile phone companies have their own wireless networking standards, including 3G (third generation) and HSPA (high-speed packet access). These systems currently offer download speeds of a few megabits per second, but in the next few years, that will increase dramatically. For example, 4G mobile networks, available as soon as 2010 in some countries, promise speeds of up to 1 Gbit/s when stationary, and 100 Mbit/s when moving at high speed – such as in a car or a train. If this sort of Internet access is also very cheap and unlimited, as we would all wish, then the era of thin clients really could begin in earnest.

In the end, all the wireless standards are likely to merge, or perhaps coexist seamlessly. The competing companies will probably force each other into price wars, in which the only winners will be consumers – eventually. This will probably happen through the introduction of hardware devices that are powerful and

Offered by mobile phone company Three, the Skypephone offers users the chance to make normal telephone calls as they would from any other mobile phone, but also allows them easy access to free Skype-to-Skype calls from their mobile phone, even when no Wi-Fi is available. The phone uses the normal mobile signal to connect users to their system, and then routes Skype calls on via the Internet. This is a sign of things to come: the merging of the various existing ways in which we use communications networks.

sophisticated enough to switch imperceptibly between the various wireless standards. Already, there is a significant trend towards 'fixed mobile convergence', in which a single device can access both Wi-Fi and mobile phone networks. An extension of this idea is 'software-defined radio', in which a device can utilise any

Within **ten years**, we really should have access **wherever we are** and **whenever we need it**, cheaply and easily

of several types of wireless network, all without the need to change mobile and wireless base stations. Current devices are dedicated to one or two standards at best, each one requiring a distinct hardware element. Ultimately, software-defined radio will probably evolve into 'cognitive radio' – a term coined by Joseph Mitola III in 2000. In cognitive radio, both the fixed antennas and the portable devices would intelligently choose whichever kind of frequency and wireless standard would be best for a given situation – be it moving or stationary, voice or data, over a localised or a wide area – and reduce the chance of interference between users. This is a mammoth task of processing that will nevertheless be possible within a few years.

Within ten years, we really should have access wherever we are and whenever we need it, cheaply and easily. However and whenever that happens, we will be living in 'pools of wireless connectivity', according to science fiction writer and technology guru Bruce Sterling. Who owns and maintains these pools, and how and how much we pay for them, is far from clear. There may be painful times ahead for consumers – who ultimately have to pay for the necessary infrastructure and new devices that can access the new services. These will be tough times, too, for those service providers who take the necessary risks and lose. In the end, however, there is little doubt that the result will be commoditised, ubiquitous, rapid and dependable access to a world of content and online services.

Up Close and Personal

One aspect of wireless networking that will find ever more applications is 'personal area networking'. This involves wireless interconnection of your personal items – such as phones, watches and headsets. It could also be called a 'last half-metre' technology that will allow your small, personal devices to be connected to the Internet at large, via your mobile phone or other more substantial device. At present, personal area networking is dominated by Bluetooth – the technology that connects mobile phones to wireless headsets, portable PCs and satellite navigation units, for example. Bluetooth is used in other ways, too: for example, there are shoes with built-in pedometers, which can send information about how far you have run wirelessly to an mp3 player; there are Bluetooth keyboards and mice for use with PCs or portable devices.

Two younger and less well-known personal area networking technologies – called Wibree and ZigBee – offer similar functionality to Bluetooth, but at reduced cost. They also use much less power, dramatically extending battery life in portable devices – to months or years. They provide connection speeds less than Bluetooth's, and are therefore useful in situations where only small amounts of information are required. They would be of little use in transferring high-quality audio or video to headphones or wireless speakers, but they are ideally suited to situations where, for example, monitoring or simple control is important.

Wibree, designed by Nokia, is the most closely related to Bluetooth – and it will be used for some similar applications, despite the fact that its transfer rate is significantly lower. There are even dual-mode Wibree-Bluetooth integrated circuits, which allow Bluetooth- and Wibree-enabled devices to interact directly. Wibree chips will appear increasingly in wirelessly controlled toys, sophisticated wireless remote controls, and 'wrist-top' devices that can connect to computers and mobile phones – displaying received emails, for example. ZigBee is likely to

The possibilities of personal area networks are endless. One Nokia concept has a wearable sensor with a strap made from solar cells, which can measure anything from personal health to the weather, and communicate data to any connected devices, such as a phone.

be used in situations in which it is only necessary to transfer small chunks of digital data. It is most likely to become the standard in home monitoring and control, because it has a greater range than Wibree (30 metres compared to 10 metres) and requires less power. A large number of sensors and controls distributed around a house will form a 'mesh network', in which data passes between and via any of the sensors – and not just to and from each sensor to a central point. Wireless control panels around the house can be used to close blinds and curtains, set alarms, and control personal video recorders and heating systems.

One area in which Wibree and ZigBee may find themselves competing for market share is medical monitoring – likely to be a huge growth area in consumer electronics over the next few years. Wearable devices that can monitor pulse, blood pressure, breathing rate, body temperature and blood glucose level are all available now, and will almost certainly become more widespread, cheaper and popular. Newborn babies wearing clothes that send vital signs to a home network could reduce the incidence of sudden infant death syndrome practically to zero. And elderly patients, or people recovering from surgery, could be monitored from a distance by doctors – reducing time patients spend at clinics and in hospitals. The growing phenomenon of remote medical monitoring is sometimes referred to as e-Health.

There are still other wireless technologies that will be vying for our attention in the coming years – in a growing menagerie of standards. Similar to ZigBee are EnOcean and Z-Wave. EnOcean is of particular interest for the future, because most EnOcean sensors in a mesh network need no batteries whatsoever: they 'harvest' energy from their environment. They can take energy from vibrations or heat – making it possible to build wireless, battery-less sensors into car tyres and implantable medical sensors, for example. In home automation, radio waves produced by a small number of powered transmitter-receivers can provide energy to power a host of battery-less sensors that form flexible mesh networks. Z-wave also forms mesh networks, and will be in direct competition with ZigBee.

Wireless USB (WUSB) is an extension of USB standards, which at short range can transfer information at the same 480 Mbit/s rate as normal, wired USB. It will probably find its greatest application in situations where USB would normally be used: connecting cameras, keyboards, monitors and printers to computers, for example. If successful, WUSB could do much to reduce the dust-collecting clutter of wires behind desks in so many homes and offices, and reduce the number of cables we need to carry with our portable devices.

There are other development efforts that aim to produce even faster wireless network speeds. In 2007, for example, scientists at the Georgia Institute of

With so many different standards and approaches to **wireless networking**, it may be some time before all our devices are **fully compatible** with **each other**

Technology achieved speeds of 15 Gbit/s over 1 metre, using a radio signal oscillating at 60 GHz. If this technology comes to market, you could download a high-definition DVD to a mobile device at a kiosk or from a friend's media player – and later, the film could stream to your high-definition television across the room with no loss of quality. This kind of scenario could be reality within five years.

A Georgia Institute of Technology researcher tests a very high-speed wireless link, sending more than 15 gigabits per second across an admittedly short distance.

REPRODUCED WITH PERMISSION OF THE GEORGIA INSTITUTE OF TECHNOLOGY

With so many different standards and approaches to wireless networking, it may be some time before all our devices are fully compatible with each other. Another major issue that must be addressed in the move towards a smoothly integrated wireless world is security. Can you be sure that those invisible signals carrying your information around will give convenience to only you? Who else might want to try to break in to your wireless digital networks? And how could they benefit if they manage to do so? Whatever the possible problems may be, the potential applications of these 'invisible' technologies are tantalising.

Physical, wired connections will still play an important role – at least in the next ten years. The first devices to make use of the wired USB 3.0 standard will appear in 2009, exchanging information at speeds of up to 4.8 gigabits per second – ten times as fast as USB 2.0. New and refurbished homes are increasingly being wired throughout with Ethernet cables, to enable the interconnection of network-enabled devices – including PVRs, games consoles as well as computers, of course. And physical connections will continue to provide the Internet's infrastructure – immensely high-volume links that span the globe. Optical fibre links will probably become the main medium supplying the 'last mile' Internet connection from service providers to homes and offices, eventually. Physical wires and fibres are very dependable digital delivery links: they provide privacy and security, along with guaranteed speeds – but they do not have the flexibility and convenience of wireless connections.

Casting the Net

By the end of 2007, around 300 million homes worldwide had broadband connections to the Internet. Uptake has been slower than one might have imagined – although millions more use broadband Internet connections at work or have dial-up access at home. Others regularly access the Internet through mobile devices. And although there are not as many broadband connections as one might imagine, the overriding trend is certainly towards a world in which high-speed, always-on connections are the norm. In 2007, Japan had the world's highest average broadband connection speed – at 45 Mbit/s. There is no reason why the rest of the developed world cannot achieve this in the next five or ten years – especially if optical fibre technology can roll out widely enough. In 2007, a 75-year-old Swedish woman became the first to receive a multi-gigabit-per-second Internet connection at home – it was provided by optical fibres that tap straight into the Internet's 'backbone'. At 40 Gbit/s, it was by far the fastest home Internet connection in the world – allowing the woman to download a DVD in less than a second. The woman's son works for a technology company, and ordinary consumers will not have access to these speeds just yet – but it shows that we can safely raise our expectations.

What wiley.com really means to a computer: 208.215.179.146.

IP Addresses

In networks that use the Internet protocol (IP), every connected device is assigned an address, made up of four numbers separated by decimal points. Each number is an eight-bit binary number – in decimal terms, this is anything from 0 to 255. Within a home, office or other local area network (LAN), there is a router. Normally, it is the router that assigns IP addresses to every device connected to the local network, and makes sure all packets of digital information arrive at their destination. Certain addresses – including those between 192.168.0.0 and 192.168.255.255 – are reserved for use in LANs. For example, the router typically has an IP address of 192.168.0.1 – so there are hundreds of thousands of routers with the same address at any one time around the world. Most of the more than four billion possible IP addresses are made available for devices connected to the Internet – and only one device can be assigned each address at any one time.

For the convenience of human Internet users, the IP addresses of web servers – computers that deliver online content – are displayed as 'domain names'. So, for example, the domain name 'wiley.com' corresponds to the IP address 208.215.179.146. Type that into a web browser's address field and you will get Wiley's homepage. It gives you an insight into the unseen workings of the digital age. Of course, in reality, that IP address is a series of binary numbers: 11010000.11010111.10110011.10010010.

Every device connected directly to the Internet at one time has a unique network address – an IP address, which allows packets of information to be routed correctly to and from it. The Internet has grown so rapidly in the past thirty years that we are currently in the fourth version of the IP addressing system. Although more than four billion addresses are available, IP version four, or IPv4, will soon prove to be unable to cope with the number of devices online – most estimates predict 2010 will be the crunch time. There are certain things that Internet providers can do to postpone the problem, but ultimately, the system needs to be improved. Thankfully, IPv6 is waiting in the wings. It makes available more than 300 billion billion billion billion addresses – more than enough for far beyond the foreseeable future. Although it was defined more than ten years ago, such a wide-ranging alteration to the way we address our online devices will not be easy to implement. It is a good thing that IPv6 will create a vast surplus of available IP addresses, because many more devices connect to the Internet to access services that were previously delivered in other ways. HSPA or 3G mobile phone networks are packet-switched, too, and all connected mobile devices have IP addresses.

Coming Together

One of the features of digital networks, including the Internet, is that all the content consists of binary numbers. Looking at a computer file in its raw binary state – as a large collection of zeros and ones – it is impossible to tell whether the file represents a text document, an image, a song or other sound file, a video, or a web page. This fact is the basis of the most important trend in consumer electronics: convergence. The functions of several separate devices can be incorporated into a single device. So, for example, in the 1970s, you could buy a cassette tape player or a record player for listening to music, a video cassette recorder for recording and watching video, a telephone for making phone calls, a

The end point of digital convergence is a kind of cheap, **universal**, **commoditised gadget** that can help you communicate and **be entertained and informed**

paper diary for recording appointments and addresses, a film camera for taking photographs, and an electronic calculator. Today, smartphones and other devices can carry out all of these functions and more. Recent games consoles also function as DVD players and Internet-ready computers. The end point of digital convergence is a kind of cheap, universal, commoditised gadget that can help you communicate and be entertained and informed. The thin client revolution and wireless connectivity everywhere really could make this science-fiction vision a reality. If it were realised, and were easy to use, such a thing would certainly threaten the dominance of the PC.

Without the ever-improving technology behind VLSI, innovative technologies such as giant magnetoresistance and fast and widespread broadband Internet connections, the digital revolution would have been a very slow affair – hardly a revolution at all. As it has turned out, we have been on a roller-coaster ride, and the still-exponential rise in storage capacity, processing power and digital camera resolution – coupled with the still-exponential fall in price per transistor and average size of VLSI components – will continue to change our lives, but at an even greater rate. Likewise, when more devices in our homes and our cars are connected together and to the Internet through even faster and more reliable connections, our attitude towards, and expectations and use of our gadgets will be quite different. We are prepared for the continuing developments: we consumers have done most of the hard work of 'becoming digital'. We moved from vinyl and cassette tapes to CD (and then, partially at least, to mp3s on PC hard disks); from film cameras to digital cameras; and from video cassettes to DVDs and personal video recorders. We will be quite unfazed if and when our film and music collections are no longer physical products on our living room shelves – held instead, perhaps, a long way away on servers, available for live streaming on thin clients wherever we are and whenever we want. Most of us are now competent digital citizens, and any shocks we now need to absorb in the continuing revolution should be less disruptive and, we hope, cheaper.

So where is this rapid and continuing miniaturisation and reduction in cost taking us – and can it continue indefinitely? The long slow death of the PC will decentralise our digital existence. What kinds of gadgets will consumers be able to buy in five or ten years? And how might we use them? In Chapter 4, I'll try to answer these questions, by looking at new approaches in the design of consumer electronics devices and new ways of interacting with them. One thing seems certain to me: we will not be relying on large, noisy, self-contained all-purpose desktop PCs for much longer.

4 Thinner and More Intelligent

The desktop PC is not dead yet, but it is coexisting more and more uneasily with a plethora of other digital devices that can store, process and display our digital information. Naturally, desktops have been fighting back, making the most of their unique talents as heavyweight Jacks-of-all-trades. For example, recent consumer models can comfortably decode, display and record television, play and back up DVDs and music. And PCs are still very popular for playing games. You can connect them relatively easily to home audio systems and televisions – if you know what you are doing – turning them into useful digital media centres. And they are still the natural choice for storing video from a camcorder, music files ripped from CDs and downloaded from the Internet, and images from digital cameras.

Perhaps most people would be unwilling to get rid of their desktop PC at present. You can almost hear their protests: 'hang on, all my stuff is on there'. But does that mean the desktop PC has been relegated to the role of glorified storage device? When that 'stuff' is stored safely elsewhere, and everything else that a PC can do can be done with other devices, the home desktop PC could disappear altogether. It will be displaced and usurped by smaller, cheaper, more flexible, portable devices. We already no longer have to sit at our desks with hands poised above keyboard and mouse to gain access to our sounds, videos, pictures and documents. When you stop and think about it, most desktop computers already look outdated.

Whatever slick, convenient, portable gadgets end up taking the place of the desktop PC – at the heart of our digital lives – they will have to live up to some of the features that at present only desktop PCs can deliver effectively. In other words, they will have to provide rapid access to large amounts of information, sufficient computing power, a constant supply of electricity and a large, clear visual display. The popular portable devices of today have small screens, small keyboards and often awkward ways of selecting information and commands – no match for the large monitor, ergonomic keyboard and mouse on the desktop. At present, they also have limited battery life. We live with these compromises for the sake of a mobile lifestyle, but if we really are about to become freed from our desktops, then the issues of how we interact with our gadgets – and access and display our information – must be addressed. There must be progress, too, on batteries. The burden is on software engineers and hardware designers to address these issues: they will need to provide us with easy-to-use, intelligent devices that can anticipate our needs. It is time to put the desktop PC out of its misery.

So in this chapter, I will look ahead to what may emerge in a new digital landscape that is not dominated by the desktop PC. I will examine some of the implications of the rise of 'thin clients' – devices that do not have massive storage capacity on-board, but which quickly access our information from more centralised, specialist storage devices. And I will explore some alternatives to the tired and cumbersome user interfaces we are used to. Finally, I will consider the current state of the art in attempts to make devices more intelligent.

The Digital Dissolve

In his book *Everyware: The Dawning Age of Ubiquitous Computing*, Adam Greenfield suggests that in the near future, information processing will 'dissolve into behaviour'. Greenfield adapted this phrase from a comment made by Naoto

Fukasawa, a Japanese product designer. Fukasawa observed that some everyday objects are so well designed that they have become almost invisible – we hardly notice we are using them. Greenfield's book examines the potential positive and negative consequences of the move towards more 'invisible' and distributed digital technology. Poignantly enough, ubiquitous computing, also called ambient computing, is often described as a 'post-desktop model' of interaction between humans and digital technology.

While we will probably use some kind of input devices to search for information and a variety of output devices to receive or view that information, most of the hardware will be quite invisible

It is not only the information processing that will be dissolving away into obscurity: the gadgets themselves will become increasingly 'embedded' and invisible. Distributed computing certainly does not mean a dozen keyboards, mice and monitors in every room. While we will probably use some kind of input devices to search for information and a variety of output devices to receive or view that information, most of the hardware will be quite invisible. It will be 'woven into the fabric of our everyday lives', to borrow a phrase from the late Mark Weiser, one of the pioneers of ubiquitous computing. In fact, some of the hardware really will be woven: wearable devices, incorporated into clothing, have been discussed and predicted for many years. This kind of electronic device is about to come into fashion – literally. For example, a team at the Massachusetts Institute of Technology's Media Lab have developed patches that can be sewn into clothes or bags, which contain a processor, memory and a wireless transmitter-receiver.

Several researchers have produced fabrics that can conduct electricity, or act as sensors of temperature, pressure or strain. Subtle, connected, wearable technology will probably play an important role in ubiquitous computing, along with portable devices of the sort we already use, and others that will doubtless be quite different.

The idea behind ubiquitous computing is not new: the concept was laid out in the late 1980s. But the technology necessary to take it out of the laboratory and into consumers' everyday lives is only now becoming readily available. Ubiquitous computing relies on a network of wireless and wired devices of all sizes, from a few millimetres up to the size of a laptop – or even smaller and even bigger. In fact, everything that could connect would be connected, at all times. Between them, the devices would provide reliable network coverage and processing power throughout a building but they could carry out specific functions, too. For example, some could be microphones or video cameras with digitising circuits on-board. These could make it possible to hold conversations and video chats while you wander about, without having to hold a handset or wear a headset. And there will be wireless connectivity beyond the home and office, too. Mobile phone networks and Wi-Fi and WIMAX networks already cover much of the territory of our cities. But what about when you step into an aeroplane or travel on an underground train – or even an overground train if it passes through patches where there is no signal? Recently, trains have been provided with Wi-Fi networks for passengers, with connections to the Internet via satellite links. And even some airlines have begun providing connectivity on-board. In a few years, these services will need to be affordable and integrated, so that information processing can dissolve when we are at large, as well as when we are at home.

Economic will and consumer enthusiasm will be necessary to make this model of distributed, dissolved devices a reality. It is unlikely that the state or a single company will own all the wireless networks that will fill the spaces in which we live and work, or the wired infrastructure that will support it. Who will? And how will we pay them? And while you might pay for the network and Internet connectivity in your home, do your devices connect effortlessly in other people's homes? Mobile phone coverage is generally much better, but again, there are certainly areas in which coverage is patchy. Mobile network infrastructure and wireless computer networks will surely converge at some point; but we may have to tread a tortuous road of expense and incompatibility before we arrive at a workable and robust situation.

The move towards a decentralised digital life can only be a relatively slow process at first. But when the necessary infrastructure and public understanding of the potential of ubiquitous computing reach critical mass, things will move more quickly. At least the devices should be able to keep up with the pace of change. Competition will continue to drive innovation, and miniaturisation and increasing computing power will still be making smaller, more powerful and cheaper devices.

An Internet of Things

The 'invisible' networking and distributed processing of ubiquitous computing can easily extend to inanimate, unattended objects, too. In our networked homes of the future, we can expect thermostats and lights to respond to us, because our presence will be announced by wireless transmitter-responders (transponders) embedded in our clothing and in our portable devices. Products tagged with radio frequency identification (RFID) chips can, if desired, make their presence felt in your fridge or store cupboard. If you are running out of toilet paper, why not have a reminder sent to you automatically when you are next in the supermarket? You may scoff, but a similar thing has already been trialled in the real world. In 2006, a system called i-Lav was installed at the headquarters of Adobe Systems in California. When soap and paper towel supplies were low, wireless devices automatically informed caretakers to replenish supplies. The same company that developed the i-Lav system has also developed a system that registers the use of water in toilets – not to catch out employees who forget to flush, but to give early warning of leaking valves in the cisterns. If all of this smacks of 'but why' then just wind forwards a few years to a time when such sensors and wireless devices really are extremely cheap and easy to connect. Then perhaps the question will become 'why not?'

The idea of inanimate objects becoming part of *ad hoc* networks is sometimes referred to as the 'Internet of Things'. One of the roles of such networks could be to monitor our environments – beyond paper towels and leaky toilets. For example, tiny sensors could transmit room temperature readings or car tyre pressures to any other device that might be geared up to receive such data. In fact, minuscule cheap sensors could even be scattered around an area almost indiscriminately. This idea is called 'smartdust', and was originated by Kristofer Pister and Randy Katz at the University of California in Berkeley. The sensors that make up smartdust would be MEMSs (microelectromechanical systems): semiconductor-based devices that are composed of more than just processing capabilities. Versions of MEMSs that could be part of an Internet of Things – that contain sensors, processing circuits, a radio transponder and a power supply – are already in development. Fully functional versions can already be made as small as a few millimetres in diameter for a few pounds each. There is no reason why within ten to fifteen years, they can't be the size of a grain of sand if necessary, and cost a fraction of a penny to manufacture. One of the problems to overcome in the development of smartdust is to invent a suitably robust and energy-efficient way for the distributed MEMSs to pass on information. What will the software be like that can make the *ad hoc* networks that deliver information to the right place? Again, much of the work towards achieving this goal is well under way.

Most of the proposed applications of smartdust are military, scientific or medical, rather than anything to do with consumer electronics. However, once the MEMSs become really cheap, and the system for monitoring them becomes sufficiently reliable, I am sure that we consumers will benefit somehow. Perhaps tiny MEMSs will be the devices that we use in our homes to help regulate temperature or measure energy efficiency, or to detect our presence and control lighting levels. It will be possible to build microphones and even cameras onto MEMSs – perhaps our homes will contain hundreds of these – dozens in each room – that can also digitise the sound when we want to speak or video conference with someone or issue voice commands. By building a tiny moveable mirror onto the surface of a MEMS chip, you could make it into a cheap, low-power projector to display information when needed.

Collectively, the different types of tiny sensor-processor-transponder devices – including MEMSs and RFID tags – are often referred to as 'motes'. In a 2005 report, the United Nations International Telecommunications Union suggested that human-operated devices – computers, mobile phones and media players, for example – may well represent the minority of traffic on the Internet of the future. If this plays out, then it seems that the massive explosion of available IP addresses that the new version of the Internet Protocol, IPv6, will bring cannot come too soon. The motes will be grabbing a large proportion of them.

Who Are You?

In the supermarket, shortly after the toilet roll holder or the bathroom cabinet back at home has reminded you to buy more toilet paper, you pay for your goods without reaching for your purse or your wallet. A chip in your mobile phone identifies you as you pass out of the shop and automatically takes the appropriate payment from your bank. As you approach your car, the driver side door unlocks. Later, when you approach a ticket barrier at a busy station carrying a heavy bag in one hand and a drink in the other, you do not have to drop everything to fumble for

RFID tags already exist. They are small, simple – but revolutionary – devices.

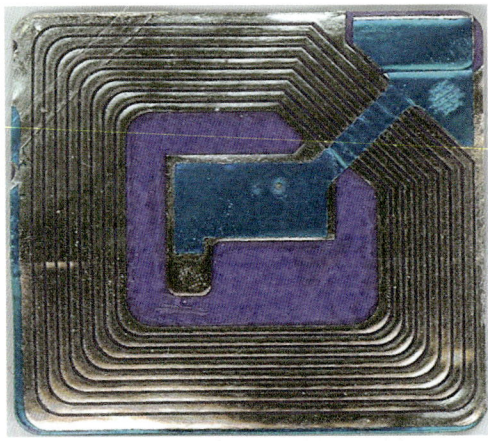

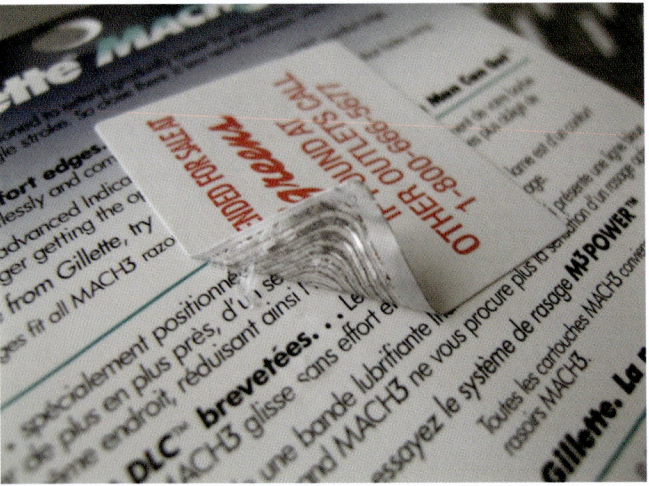

Being correctly identified brings the
promise of convenience and security:
you will be able to access your content, but
you will not be able to choose who else has
access to it

your paper ticket or swipe card: the barrier opens for you as you stride towards it. It is clear just from these few examples that an important aspect of ubiquitous computing is identification. Being correctly identified brings the promise of convenience and security: you will be able to access your content, but you will not be able to choose who else has access to it. Of course, this is not necessarily true when it comes to digital rights management: you will not be able to give others access to music or video they have not purchased. Anyway, at the bank, at the office, at the airport and online – in fact, anywhere that identification is important or desirable – electronic technologies that can quickly, correctly and securely identify us could bring tremendous advantages. There would be no more paper tickets, little use for cash or even credit cards, and potentially, no need to worry about losing your keys.

Identification systems can utilise electronics or biometrics as the identifying feature, depending upon the circumstances and on the level of security needed. Electronic systems include RFID chips in keys or ID tags – or even chips embedded under the skin – and 'near-field communication' chips in mobile phones and other portable devices or smart cards. Biometric identification technologies include fingerprint and iris readers and face and voiceprint recognition systems. All of these already exist, including fingerprint readers built into portable PCs – but in a world of ubiquitous computing, their importance and their power will grow.

No one can deny the potential convenience that personalisation by identification could bring in many situations. And, unlike current methods of identification – such as passwords and paper documents – fingerprints, faces and even embedded RFID chips are hard to lose, steal or forget. But however convenient electronic identification may be, many people have strong concerns about it. In some quarters, the reaction goes as far as rage and disobedience.

With your bank account, your social security records, and perhaps entry to your car and house all tied in to, say, your fingerprints, an identity thief only has to steal a single piece of information to 'become' you. Of course, fingerprints are difficult to steal – although biometric security systems must be designed to reject copies of people's fingerprints, such as casts or photographs. Just as secure as a fingerprint is a voiceprint: a unique collection of identifiers that stem from the characteristics of a person's vocal tract and the sound of his or her voice. After extensive trials and proof of concept, the bank ABN Amro in the Netherlands became the first to use voiceprints as a way to identify its customers. When customers call the bank, the bank's system can verify their identity to a very high degree of accuracy. It works whether the customer is using a landline or a mobile phone, and even works when the caller has a cold. For now, a customer has to answer a predetermined question, but at some time in the future, simply phoning up and saying your name should be enough to identify you.

Electronic means of identification are easier to steal: any radio transponder device can be scanned and its output replicated. You can find instructions on how to clone an RFID chip on many web pages. More sinister perhaps, and certainly more controversial, is the possibility of increased surveillance and the potential loss of privacy. Already, some ID tags can record entry and exit times and work out where you are during work hours. RFID tags attached to products – perhaps unknown to the consumer – could reveal that person's whereabouts and activities, especially as records of transactions can link credit card numbers to the RFID chip attached to the product. For people already under surveillance – including, for example, political activists – there will be no escape. In fact, some opponents of RFID call the devices 'spychips'.

In 2004, the US Food and Drug Administration (FDA) approved a small glass-encased microchip that is implanted just beneath the skin in the right arm. This is the same kind of device as 'pet passports', which have been in use for some years. The chip responds to a nearby source of radio waves of a particular frequency by transmitting a sixteen-bit number. It draws its power from the radio waves, so it needs no internal power supply. Called VeriChip, the device has been the source of considerable controversy. The stated uses for the device include matching a patient with his or her medical records – which could be useful in an emergency situation as well as convenient in hospital and at the doctor's surgery. A similar chip is under development that can monitor and transmit measurements of blood glucose concentration – a device having particular resonance with sufferers of diabetes. By the end of 2007, more than two thousand people had received a VeriChip. But the company that markets them, VeriChip Corporation, estimates that eventually 45 million Americans will have one. Various companies in the USA have suggested the compulsory 'tagging' of their employees, though legislation has so far blocked this proposal. One further concern is the possibility that frequent scanning of the device could cause cancers. A few studies in mice in the late 1990s suggested that this could be the case, but there is no evidence to date in larger animals. VeriChip and the FDA are confident that there are no tangible risks.

When and if people are convinced they can trust technologies that routinely identify us, and that the threats against their privacy are small, we will move one step closer to ubiquitous computing proper. Actually, we will almost certainly move that way anyway, whether or not people are convinced of the safety and reliability of these identification technologies. It is possible to have a kind of ubiquitous computing without frequent identification, but it would be less convenient, and less powerful. Will we be happy to relinquish yet more of our privacy for the sake of digital convenience?

All of this talk of inanimate objects connecting to each other – the Internet of Things – seems to be a bit removed from consumer electronics, since it is more concerned with sensors and surveillance than information and entertainment. However, motes of every flavour will provide a personalised, intelligent, contextualised and more natural interface between ourselves and the digital domain in a world of ubiquitous computing. But how will we interact with distributed devices? Where will we type or speak or click or touch?

Interfacing the Future

Computer scientist Bruce Ediger is often credited with saying that 'the only interface that is intuitive is the nipple; after that, it is all learned' – although no one seems sure who actually said it first. The point is clear, however: interacting with our gadgets does not come naturally. It is something we have to learn. Although many of us normally choose not to read through complicated manuals when we buy a gadget, doing so would probably highlight several features of our devices of which we would remain unaware. Life is just too short. Having to read manuals or to learn new skills in order to get the most from your devices is symptomatic of the fact that Ediger is right – for now at least. The move towards ubiquitous computing will dramatically reduce the number of discrete, fixed or mobile devices we possess – perhaps eventually to zero. It will necessitate the invention of more natural ways of interacting with and accessing digital information, completely at odds with the PC-based model with which we have been familiar. Here is a chance for software engineers and hardware designers to prove Ediger wrong.

Interacting with **our gadgets** does not **come naturally**. It is something we have to **learn**

The way information is organised and represented visually on our computer monitors today is called the graphical user interface, or GUI (pronounced 'gooey'). It is a very clever invention, because it masks the fact that all you are actually doing when you click a mouse button is sending streams of binary digits to the CPU, to initiate other streams of binary digits. Making sense of all those zeros and ones in such a simple way is quite a feat. We are so used to the idea that files reside in sub-folders that reside in folders that it is hard to accept the fact that this is not actually the way they are arranged on a hard disk. Many of us know our way around small mobile phone keypads so well that we can write a long text quickly and with hardly a glance at the phone or its keypad. And reaching for the mouse and typing on a keyboard has also become second nature; it almost feels intuitive – but of course, it, too, is learned. One key feature of the GUI approach and mobile phones is the use of menus and sub-menus. In a word processing program, you will find commands to change the level of magnification of your document in the 'View' menu. Where do you look for the 'Print' command? In the 'File' menu, since it applies to the document, or file, as a whole.

Finding your way around menus is a transferable skill. Not only most applications, but also portable devices tend to use stripped down interfaces that nevertheless work in a similar, menu-driven way. If you want to change the equalisation setting on your mp3 player, you will find them in the 'Settings' menu, but you will not find any songs or playlists there. Instead of a mouse, portable devices typically have scroll wheels or small joysticks to find and access menu-based commands. Games consoles normally offer visual choices with which the gamer can interact, using buttons on a game pad.

The Graphical User Interface

In the 1960s, and through much of the 1970s, the only way people could communicate their instructions in real time to a computer was via text-based input. This approach is called the command line interface. Before Microsoft introduced its Windows operating system, it relied upon the Disk Operating System (MS-DOS), which had a command line interface. You can still use it – and Apple's Mac OS, Linux and other operating systems have similar command line interfaces, too. However, the average user is never going to learn a collection of specific commands, and the syntax that goes along with them, to interact with his or her computer in that way. In 1978, researchers at the Xerox Palo Alto Research Centre (PARC), California, developed the idea of the 'graphical user interface', or GUI. Apple introduced the GUI into consumer machines in 1984, and Microsoft followed suit the following year with its first Windows software. And the GUI has dominated personal computers ever since. The Xerox researchers designed a system that still persists today: it 'opened' files and folders in moveable windows; it used icons to represent different types of files and folders; commands were organised into categorised visual menus, and all of these features were accessed using a mouse.

While this menu-driven and GUI-based approach is well suited to personal computers and some portable devices, it does have its pitfalls. Switching from keyboard to mouse and back again can be annoying – although there are hundreds of 'keyboard short cuts' that you can use to avoid having to select commands from menus using the mouse. Of course, most people only learn a few of the most important short cuts. Drilling down through sub-folder after sub-folder and sub-menu after sub-menu can be tedious. 'Right-clicking' on a Windows- or Linux-based computer or 'Control-clicking' on an Apple Mac computer can help to reduce the annoyance, by bringing up contextualised menus that contain only the commands relevant to a particular situation. While this approach eases the burden a little, it is still firmly rooted in the existing model of how we organise our digital information. But there has to be another way: a radical departure from the top-down hierarchical structures we are used to.

In tomorrow's world, information will have to be more 'fluid'; it will have to exist in many places at the same time and be manipulated and shared effortlessly and appropriately. In his book *Everything is Miscellaneous*, David Weinberger provides an excellent manifesto for dealing with this superfluidity of information. Weinberger points out that historically, we have tended to organise information neatly into categories. For example, in library cataloguing systems, textbooks and office filing systems, information can normally only exist in one place. We have had categorisation thrust upon us, but in the information age, this approach is no longer relevant or necessary. Much of what we do falls into many different categories, potentially an infinite number – or perhaps none, depending upon how you look at it. Should the photograph you took last week be in a folder that contains all the photos you took last week? Or should it be filed away in a folder that contains all the photographs you have taken of your family? Where you put it has consequences for how easy it is to find it again. But in the digital domain, that photograph can 'exist' in all of those places and none of them if we move away from the folder-based structures we are used to.

To manage and make the most of this emerging digital landscape, we need a breakthrough as profound as the introduction of the GUI – or perhaps many such breakthroughs. In the past few years we have already seen, if not breakthroughs, real breakaways from the top-down, menu-driven, inflexible interfaces we are used to. The World Wide Web has led the way. Originally, it simply extended the way information is organised on computers across a network, but even right from the beginning, it offered something different. Hyperlinks – a central feature of the Web – connect to related pieces of information that a user can instantly access if he or she wishes. Now, with the coming of the Web 2.0 – that revolution in 'social computing' and user-generated content – we have another very important way of undermining the strict organisation of information: 'tagging'. By associating simple words – the tags – with a file or a piece of information, that information can exist in many categories at once, and is findable in myriad ways. This is the equivalent of finding a particular CD in several different display shelves in a music shop, each one covering a different category. In the physical world, this would mean that record shops would have to be enormous. On the Web, you can find what you want – and potentially discover much more of interest – by searching or browsing using tags.

A tag cloud that relates to the Web 2.0.

What kinds of device might we use to access and share our information? What will the interfaces between ourselves and our digital information be like?

Hands On

So the parts necessary to make ubiquitous computing a reality are coming together: networking everywhere (when service providers can coordinate and integrate better and more cheaply), identification technology (although it remains controversial) and freeing our information from its prison of categorisation. But what kinds of device might we use to access and share our information? What will the interfaces between ourselves and our digital information be like?

Manufacturers have already begun moving away from the traditional methods of interacting with our gadgets. They are thinking beyond the standard keyboard or keypad, mouse or scroll button and passive screen. Nowhere is this more apparent than in games consoles. In 2003, for example, Sony introduced a camera called the EyeToy on its PlayStation 2. Although some games simply incorporate the image of the player into the gameplay, the most important feature of the EyeToy is that users can interact directly with a game by waving their hands around in front of the camera – gesture control. Their hands – or rather the image of their hands – become the equivalent of a mouse pointer, activating certain commands and choices on-screen. The EyeToy caused quite a stir in the gaming community, and did much to attract a wider audience to PlayStation 2 console.

Inside the Virtusphere, virtual reality gamers can walk or run around, thanks to gimbals on which the sphere is mounted. This adds another level of realism to the game.

Moving Games

Virtual reality has been in development for decades. But making a convincing virtual representation of a three-dimensional space requires a good deal of processing power, so it is only recently beginning to realise its potential. Modern computer games consoles create extremely detailed virtual three-dimensional environments, but the gamers can only normally experience it in two dimensions – on the television screen or computer monitor. Virtual reality headsets and gloves enable gamers to interact with this kind of environment directly – as if they were there. One problem with virtual reality headsets in the past has been the restriction in movement: the gamer is stuck motionless in the real world while he or she is moving in the virtual one. An innovative solution to this is found in Virtusphere, a closed sphere supported on gimbals – balls that allow the sphere to turn in any direction. The gamer stands inside, and can walk or run in the sphere, which turns freely, feeding back its motion to the games console. This enables the gamer to move inside the virtual world by walking or running in the real world – but without moving from the spot.

In a different solution to the same problem, gamers move around a real space, wearing virtual reality headsets – or even just carrying a mobile phone. In their headsets, they can see the real world with a virtual gaming space overlaid. With a phone, the screen displays what the phone's camera can see, but with information superimposed. The real and virtual worlds are coordinated because the gamers also wear satellite navigation device and motion sensors. This idea is called 'augmented reality' or 'mixed reality' gaming.

The innovative controllers of Nintendo's Wii games console did even more to extend the demographic of 'gamers' beyond the traditional young, technologically savvy males. The Wii Remote, sometimes referred to as the 'Wiimote', has a built-in accelerometer – a device that registers the speed and direction of movement in all three dimensions. The accelerometer in the Wiimote is a MEMS device: it is built on a single semiconductor chip. The controller also detects the direction in which it is being held, via a light sensor. The sensor picks up infrared radiation from LEDs (light-emitting diodes) in a 'sensor bar' that sits on top of the television to which the Wii console is attached. Finally, the controller can vibrate to give 'force feedback' in certain games. Altogether, the effect is a fairly intuitive interaction: you play tennis by swinging the Wiimote with one hand, for example, rather than using your thumbs on a game pad as you have to do on most consoles.

Small, cheap, solid-state accelerometers are used in other devices besides the Wiimote. The iPhone and iPod Touch use them to orient the screen in landscape or portrait format depending on how the devices themselves are oriented. Tablet PCs have been doing this for some years, again by virtue of accelerometers. Several smartphones do the same, and some also incorporate a 'shake' function that allows the user to skip to the next track while listening to music.

Another function common to many Tablet PCs, to the iPhone and the iPod Touch, and to some other handheld devices is touch screen capability. Using fingers or a stylus to interact directly with a display is certainly not new – it became popular in the 1970s, but in a limited range of applications that did not really venture into the consumer market. But in the past few years, it has begun to make its presence felt in consumer electronics. Building a touch screen into a device allows designers to reduce the number of physical buttons a device has to have – often to just a power button – and this means the size of the screen itself can be maximised. The recent touch screen revolution really began with in-car satellite navigation systems around the turn of the century. Today, touch screens have become more common; you can even buy small devices that you clip onto your existing computer monitor that transform it into a touch screen monitor. There are many different ways of sensing where on a screen someone's finger or a stylus is touching. For example, some screens simply sense pressure, others detect changes in electrical parameters such as resistance or capacitance, while still others create a field of ultrasound across the screen surface and detect how and where fingers disturb it.

Of course, a touch screen alone does not represent a radical departure from the existing model of how we interact with our devices. It is just a selector device – a replacement for a mouse, a joystick, a scroll wheel or buttons. But extend the ability, so that a device can sense two or more points of contact, and things do start to evolve. This approach, called 'multi-touch' technology, is already providing a new user experience in several gadgets. The 'pinch' function of the iPhone and iPod Touch, with which you can use two fingers to zoom in and out of photographs and web pages, and the flick function, with which you can scroll quickly through a long list or search quickly through songs or albums with a flick of a finger, really do enhance the user experience. And they feel more intuitive than scrolling a mouse wheel or clicking 'zoom' or 'next'.

In 2006, computer scientist Jefferson Han demonstrated some truly remarkable software that makes good use of multi-touch technology. A video of his talk quickly became popular on YouTube, and inspired many people to have expectations beyond the traditional GUI model. In his talk, Han showed how he could move, pan, zoom and rotate digital photos spread across the display screen using up to ten fingers. He interacted with several other applications, including animation software and a three-dimensional atlas program. Each was completely intuitive – there were no menus, and in each case, simple multiple-finger gestures enabled full interactivity and control over the information displayed. A certain touch gesture brought up a virtual keyboard, so that Han could enter text. Multi-touch interfaces were first developed in the 1980s, but the kind of applications that Han and his colleagues have developed require huge amounts of computing power that were not readily available back then. In his talk, Han said that he cringed at the thought that a whole new generation is about to learn how to use the standard 'mouse, windows and keyboard' model of computing, rather than the much more intuitive multi-touch technology.

In a hotel reception, Microsoft's Surface can display interactive maps and over dinner you can even order food and drinks by interacting with your table

Many other people are working on multi-touch devices along the same lines as Han. Of particular note is Microsoft's system called Surface Computing, which was announced in 2007. It is basically a Windows computer with added software to enable the multi-touch facility. The screen itself is set horizontally, and is embedded in a table that contains the hardware necessary to detect objects placed on the surface. It has the same kind of functionality as Han's device, but you can also transfer photos and other information wirelessly and effortlessly to and from digital cameras, mobile phones or other devices, as long as they have wireless capabilities. Philips have produced a similar product, called Entertaible, which is dedicated to table-top games. For now, these devices are designed for public spaces, such as shops and hotels and bars. In a hotel reception, Microsoft's Surface can display interactive maps and over dinner you can even order food and drinks by interacting with your table. The Entertaible's 32-inch horizontal screen allows for intuitive multiplayer interaction, ideally suited to pubs and clubs. Even if you wanted one in your home, they remain prohibitively expensive for most consumers – but within a few years, they are sure to become more affordable and more general purpose, and will no doubt do much to challenge the PC's dominance.

Eventually, multi-touch hardware and software will become commoditised, appearing in a wide range of non-branded products, rather than a few branded ones. It will no longer stand out as something new. And it will be very affordable. But while we will probably see several multi-touch screens around our house,

Another project that follows this intuitive, multi-touch, surface-based idea is the 'reacTable'. This table-top-sized device is a collaborative musical instrument. Various blocks can be placed on the table to initiate different sounds and effects. Rotating the blocks and touching the table around them controls volume, pitch and tempo. Moving the blocks relative to each other changes the effects and associations between the different sound generators. Icelandic singer Björk became the first person to use a reacTable live on stage, in 2007.

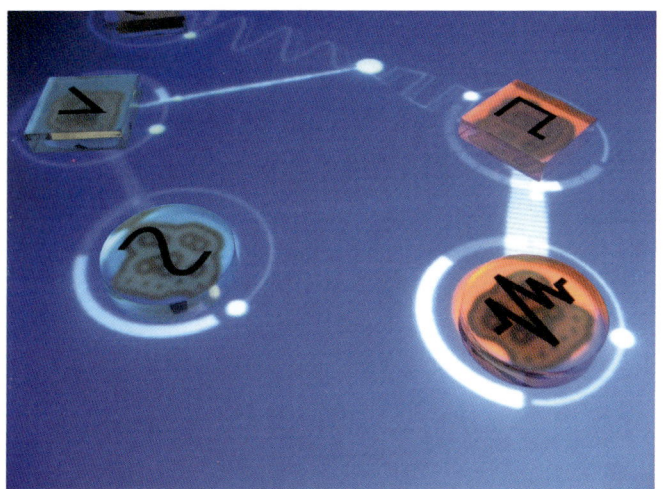

projected images may be more common. We could have very small, cheap devices in mobile phones that can project displays onto walls or other surfaces. How would you interact with such images? Well, you could use your voice or a wearable controller. Alternatively, you might just be able to control a pointer on the display by moving your hands in the air. If you have seen the film *Minority Report*, you will probably remember the scene where the police chief flicks quickly through many screens of information simply by waving his hands in the air. There are several researchers currently working on making this a workable solution in the real world. At present, most have produced systems in which the user has to wear special reflective gloves, and two cameras must be mounted in the room to detect the user's hand movements. As I write this, there are rumours that the next version of the Wii may not have handheld remote controllers at all, sensing instead physical movements of the hand and body to control the game. If that rumour turns out to be false, I am sure it will not be long before a games console – and eventually many non-portable devices – are controlled, intuitively, in this way.

Gesture-based interfaces like these will be increasingly important, I am sure, and will help to make desktop PCs a thing of the past. I have already mentioned another way of interacting with our digital devices without mice and keyboards, and without moving our hands at all. We can talk to them – and if we are lucky, they might just talk back. Today, it is very common for mobile phones to have voice dialling functions: just speak a name into your phone's microphone and the phone will dial that person's number (if you have previously stored it in the phone). This involves speech recognition rather than voiceprint recognition.

Talking Clever

It is not only mobile phones that make use of speech recognition technology. On PCs, both Windows Vista and Mac OS X have sophisticated speech functions built in – and this technology can be added to other operating systems. You can issue any of a number of spoken commands and teach the computer new ones; speech functions on a PC or mobile phone also allow you to dictate documents or messages. In 2007, Ford and Microsoft introduced an impressive in-car system that uses voice commands to control music and make and take phone calls almost completely hands-free. The system, called Sync, also uses Bluetooth to connect audio devices and mobile phones in the car: it is like a car-based precursor of ubiquitous computing.

Speech recognition software is able to pick out spoken words by identifying the individual sound elements of speech, called 'phonemes'. All of the words in the English language are constructed from just 42 phonemes, each with a distinctive sound. These sounds make up the phonetic alphabet that is used to denote pronunciation in dictionaries. Phonemes can be singled out by analysing the frequencies of the sound, and the variations in its volume, and comparing these with a database of previously recorded versions. Once the phonemes have been identified with a fair degree of certainty, speech recognition software makes use of a complicated statistical procedure to infer what words were 'probably' spoken. For this, it uses the fact that some words go together more commonly than others. So, while the following combinations of words sound very similar, it is fairly easy to guess which one would more likely be spoken: 'You were in Beirut' and 'Ewe whirr inn bay root'.

The whole process requires significant amounts of processing, and effective, real-time speech recognition has only been possible in the past ten years. Success depends upon the level of ambient noise and speed of speech, but in ideal conditions, even an ordinary consumer level PC can achieve 99% accuracy. One of the most important centres of excellence in speech recognition is Carnegie Mellon University (CMU). In 2005, its researchers produced application-specific integrated circuits (ASICs) that could take some of the burden off a computer's main CPU chip, in the same way as PCs have separate chips that carry out much of the graphics processing. By shifting some of the processing from software to hardware, the CMU researchers were able to achieve better than real-time speech recognition. This does not mean that the system could work out what was being said before a sentence was finished: it means that the processing of a recorded phrase could be carried out in less time than the phrase took to say. The CMU researchers suggested that by shifting the speech recognition process further into dedicated hardware, future systems would be able to carry out excellent results in the equivalent of one-thousandth of real time.

Silent Commands

Researchers at NASA are working on a system that can recognise spoken words that are not even spoken out loud. When we mouth words silently, we utilise the same muscles in the tongue and the throat as when we speak out loud. The technology involves button-sized sensors that attach to a person's throat and detect the nerve signals produced during this 'subvocalisation'. Software analyses the nerve impulses and can work out what words are being spoken. The researchers are working on a version that can work through clothes: so, sensors attached to a shirt collar, for example, could pick up spoken commands or dictation in, say, noisy environments or where the user has to be quiet.

NASA researchers can give voice commands without actually speaking.

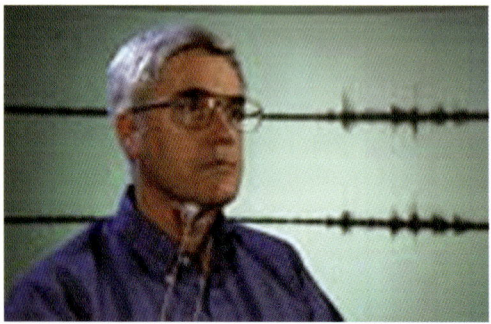

One of the other main research efforts at CMU is to produce a wearable device that can translate one language to another in real time. This is a significant

A **simultaneous translation device** would hold massive appeal to consumers in a number of situations, not least **tourism** and international business relations

challenge, because not only must their device recognise words, it must also make some kind of sense of sentences, before reconstructing those sentences in another language. A simultaneous translation device would hold massive appeal to consumers in a number of situations, not least tourism and international business relations. It is likely that a reliable model will be available within ten years – though even then, its capabilities may be limited.

It is a real achievement to make devices that can receive spoken commands and take dictation, and that can produce real-time machine translation of spoken words. But it would be even more useful, and would provide a completely intuitive human–computer interface, if the computer could answer back intelligently. Using natural language and holding two-way conversations with your computer or other device feels like a long way off. Similarly, although the new interfaces already described offer new, more intuitive ways of interacting with digital devices, they are still limited in what they can do, and are not of themselves a completely radical departure from what we already use.

To revolutionise all these forms of interaction, significant progress is needed in perhaps the most challenging area of digital technology: artificial intelligence. With a truly intelligent system, the user does not have to find the necessary command by clicking his or her way through menus. Such a system can anticipate the user's needs in any given situation, and be able to interpret casual gestures and imperfect commands or requests.

Intelligent Machines

Although computers and other digital devices sometimes appear intelligent, there is no real thinking or anticipation going on inside their 'electronic brains'. For the moment at least, human brains and CPUs are worlds apart. In most practical examples of artificial intelligence (AI) in consumer electronics, it is the software designers that do the thinking and anticipating while they write the programs that make their devices appear intelligent. They build the relevant responses and actions into their programs or employ mathematical functions called algorithms: this approach is called 'top-down AI'.

When designing a top-down AI system, programmers consider intelligence as a mathematical or computational problem. Can they produce an appropriate, seemingly intelligent, output from a given input? Top-down researchers do not concern themselves with modelling how the human brain works: the brain may as well be a mysterious 'black box'. In a 'closed system', with limited inputs and outputs, this approach can achieve impressive results. Inside a satellite navigation unit, for example, top-down AI can plan routes by following strict algorithms, which can even review the route if it is alerted to a traffic jam or a road block. This appears intelligent – 'isn't it clever?' we all say – but of course, the device is blindly following orders, looking through all the options and choosing the best one according to a program. A very simple top-down program can win every time at noughts and crosses – but it would have to be reprogrammed if you wanted to try out a new variation of the game. We humans, on the other hand, could adapt to a new rule – without the need for reprogramming. Very powerful computers can even beat grandmasters at chess – something that very, very few humans are able to do. In 1997, IBM's Deep Blue computer became the first machine to beat a reigning world champion chess player. The computer won a six-game match against Garry Kasparov. Three games were drawn, and Deep Blue won two of the other three. Chess

It is telling that a computer that **can beat a grandmaster at chess** cannot **converse** as well as a three-year-old

computers can process very quickly all the possible outcomes many moves ahead, assessing at each potential stage of the future game whether they would be in a good or a bad situation. Even grandmasters are not able to think of all the possible moves more than a few moves ahead. But they use different tactics that can pay off against all but the world's most powerful supercomputers: we human beings have capabilities that are impossible to reproduce in a top-down AI system. We are naturally creative, we can 'think outside the box' and we can feel emotions, and we don't have to be programmed. It is telling that a computer that can beat a grandmaster at chess cannot converse as well as a three-year-old.

In top-down AI, ideas and thoughts are effectively represented as binary numbers, just as characters, images and sounds are in any digital device. A computer manipulates 'ideas' and 'thoughts' according to rules. The top-down approach to, say, recognising a model of car would be to compare angles and shapes with an extensive database of car statistics. By making a database large enough, and by using enormous amounts of processing power and memory, a computer running top-down AI software can make seemingly intelligent inferences and judgements. Of course, bring in a new car model, and the database will be useless, temporarily. One significant and long-running top-down AI project is Cyc, a huge database of human knowledge that has been growing since 1984. Run by a small company called Cycorp, the Cyc database now contains close to half a million 'concepts' – elements of the world, such as the names of colours, ideas and objects – and more than three million 'assertions', simple phrases such as 'Dogs bark' and 'Trees have roots'. Software that works with Cyc uses 'knowledge engineering' to make logical inferences to connect concepts. The database does not learn new concepts and assertions automatically: these have to be entered manually. Although the project has been running for more than twenty years, researchers suggest that the database is only a tiny fraction of the size necessary before it could rival humans' 'common sense', let alone something approaching human-like intelligence.

Top-down AI can be very effective at analysing sentences – a process called parsing. All properly formed sentences consist of identifiable phrases. So, by knowing that 'Mike' and 'the table' are nouns and 'sees' is a verb, a computer can successfully parse the sentence 'Mike sees the table.' Of course, the computer will not have any real understanding of that sentence: it will not 'internalise' the meaning or experience any emotions or associations from it. But it may be able to produce seemingly intelligent responses to well-formed questions. For example, the question 'What time is the next train to Leeds?' can be parsed, and if the necessary knowledge is available, an answer can be constructed. But such a system cannot hold a sensible conversation. It is clear that top-down AI is nothing like thinking. Of course, it is not necessary for a computer to think in the same way as we do to act intelligently. But genuine intelligence is about responding appropriately to the world around you. In any real world situation, where the world is messy and unpredictable, any number of rules and inferences will not help you.

The alternative to top-down AI is 'bottom-up AI'. This approach simulates the way our brains work more closely, by modelling how individual neurones behave and interact. Neurones are nerve cells, and the brain is composed of around 100 billion of them; each neurone is connected to up to a thousand or so others, so that the brain contains trillions of connections. The highest density of interconnection is found in the cerebral cortex – that wrinkled part at the top of the brain. This is where we generate our perception of the world, where memories are laid down, and where most of our higher faculties are generated – including speech and social skills.

Sensory neurones deliver electrical signals from nerve endings – light-sensitive ones in our eyes and pressure-sensitive ones in our ears and in our skin, for example. This input is spread across thousands or millions of neurones, and each one responds by 'firing' a train of electric pulses or 'not firing' any pulses. Because of the interconnectedness inside the cortex, this leads to an insanely chaotic, unpredictable pattern of electrical stimulation inside your head – but one that can 'process' all the sensory information, associate it with memories, and produce organised, relevant outputs. For example, motor neurones carry orderly and coordinated signals to our muscles, including those in the face, throat and chest that make us speak.

One of the most important and amazing faculties of the interconnected networks of neurones inside the brain is the ability to learn. When we learn, new connections form or existing connections change their patterns of response. In other words, the brain is constantly rewiring itself. This is why the top-down approach to AI seems like a non-starter to many researchers for all but very closed systems.

The understanding of how neurones work stems from research carried out in the 1920s, although science has come a long way since then. It is possible – in fact, it is fairly easy – to model the behaviour of individual neurones using a computer program. Bottom-up AI basically involves modelling many neurones together within the same program, and including the connections between them. The result is called an artificial neural network (ANN), and it can produce some very interesting results. For example, ANNs can learn to recognise faces or patterns in data. A simple robot with an ANN that is given simple goals but little programming can learn to walk or produce cooperative interactions with other, similar robots.

These behaviours 'emerge' from the behaviour of the ANNs – they are not specifically programmed. The ultimate hope of bottom-up researchers is that intelligence itself may 'emerge' from a sufficiently complex artificial neural network.

If software engineers and **hardware** designers do manage to **instil real intelligence in our gadgets**, you can be sure it will require **huge amounts** of processing **power**

If software engineers and hardware designers do manage to instil real intelligence in our gadgets, you can be sure it will require huge amounts of processing power. If Moore's Law continues, perhaps that computing power will be available in the very near future. And just as the dedicated speech recognition chips being developed at Carnegie Mellon University will probably develop into plug-ins for a number of different systems, so a variety of AI applications might one day be available on stand-alone chips.

Artificial intelligence is sure to play an increasingly important role in our digital lives. For example, non-human players in computer games will be more realistic and unpredictable, and may be able to learn more and quickly about our approach to the game. Cars will be able to parallel park better than any human can, and will even drive themselves. Robots already make use of sophisticated artificial intelligence; many different products are already available to consumers, although they are limited to toys and simple maintenance tasks, such as house cleaning and grass cutting.

Helping Around the Home

Small domestic robots are already available for limited, repetitive chores – tasks that still need some intelligence if an autonomous device is to carry them out, with little or no human input. The Roomba is a robot vacuum cleaner, while the Scooba is a floor cleaner. Both can do their jobs effectively while you are out at work – and they return to their charging stations when their batteries are running low. Similar robots are available for cleaning swimming pools, mowing lawns and clearing leaves from gutters. Stepwise improvements in this kind of approach, along with increased processing power and better artificial intelligence, will see a widening range of such consumer devices.

Our personal gadgets will have to become more intelligent in order to cope with the fluidity of information – as well as the sheer amounts of it. They will help us to filter out what we do not need to know, to summarise large amounts of information we would want to know, and to alert us to anything of potential interest to us. In other words, new interfaces that give us more flexible access to information will have to do some of the thinking for us. Electronic 'intelligent agents' will take the drudgery out of searching for and collating information. They might scan thousands of web pages, looking for certain information, such as product prices or the latest news. They will have the ability to learn about our likes and dislikes and our expectations. In fact, some online shopping sites already make use of AI to recommend new products based on a shopper's previous purchases. At present, if you want to know a piece of information – say, 'Who was the actor who played Rambo?' or 'What is the time difference between Tokyo and New York?', you can find it quickly and easily, but you have to select search terms and filter through the results in a search engine. Within ten years, you may only have to ask your digital devices a question, and they will do the work for you, and speak the answer back. Booking flights, finding out the best prices of something you want to buy, or researching your family history could become much more convenient.

For artificial intelligence to make a real impact on the way we use the Web, some web-based information will need to be formatted in a way that software agents can 'make sense' of it. Software engineers have already begun designing such formats, which are set to become the basis of the 'semantic web' – part of the vision of the next generation World Wide Web, dubbed 'Web 3.0'. Some futurists predict that the Web will develop into a vast and changing database, much of which can only be read by computers, which will then serve up the relevant information in the most relevant way. This potential development only serves to emphasise the declining importance of the current GUI-based, menu-driven model of interaction with digital devices.

Autonomous Vehicles

One common aim of artificial intelligence is to produce vehicles that can navigate and drive or pilot themselves without human intervention. Every year since 2004, a competition called the DARPA Grand Challenge has brought together the most promising teams working on self-driving cars in a race held in California. To win, the robot cars have to complete a course in the shortest time and with the fewest mistakes. The cars taking part in the challenge are fitted with cameras, laser guidance, satellite navigation and powerful on-board computers running state-of-the-art AI software. In 2007, the course was a testing urban environment: there were traffic lights, and the cars had to drive around together, anticipating each other's movements. Although the contest aims to stimulate the development of autonomous vehicles for the battlefield (since DARPA is the US Defense Advanced Research Projects Agency), it brings the dream of consumer-level self-driving cars ever closer.

A post-desktop future of consumer electronics is coming. With the demise of the PC, our clunky, menu-driven devices will also be replaced by intelligent, intuitive devices that will give us free and easy access to the information we want and wherever we want. While many of the developments I have discussed in this chapter are attainable with today's technology, or with near-future improvements on it, some will rely on completely innovative technologies still in the laboratory – which I will explore in the next chapter.

Images from the 2007 DARPA Grand Challenge for autonomous vehicles.

5 No Detour Ahead

So here we are in the midst of a quickening revolution in the way we record, manipulate, store and share information. The integrated circuits that sit at the heart of our consumer electronics devices are becoming cheaper and more powerful at a tremendous rate. And the transistors that form the basis of their operation are becoming smaller and faster equally rapidly. These changes are encapsulated in Moore's Law, Gordon Moore's famous observation that the number of transistors you can fit on an integrated circuit doubles about every couple of years.

If manufacturers can keep up the pace of development in consumer electronics, then fifteen years or so from now we will have chips packed with transistors not much bigger than individual atoms that can switch on and off a trillion times every second. They will use much less power than today's processors and memory chips, and they will cost next to nothing. If that happens, then we can look forward to portable devices with capabilities beyond today's supercomputers. Progress in consumer electronics is not restricted to CPU chips: we can expect storage devices that can hold a lifetime's information and still fit easily into the palm of your hand, a whole new range of thin, flexible, low-power displays and batteries that never need charging.

Fifteen years or so from **now** we will have chips **packed with transistors** not much bigger than individual atoms that **can switch on and off a trillion** times **every second**

But what if the consumer electronics industry falters on the road towards tomorrow? What if, five years down the line, miniaturisation has gone as far as it can? The Internet would continue to develop, and the prices of existing technologies may fall to next to nothing, but we would be stuck. Part of me doesn't think that would be so bad – it would give us as chance to find a way to use electronics sustainably, and in the best ways possible, rather than pursuing progress for progress' sake. But at the same time, I am keen to see what kinds of technology might be possible if chips, hard disks and display screens do continue to improve and become cheaper. So, what is it to be? Can manufacturers continue packing ever more transistors onto their silicon chips? Will our digital devices ever be so cheap and ubiquitous that they dissolve into the backgrounds of our lives? Will they be able to make better batteries and flexible display screens? Will we ever have enough computing power in a cheap, portable device to make it truly intelligent?

It turns out that there is trouble ahead; to understand what kind of trouble, consider the following scenario. A sprinter has set himself a challenge. He has to improve on his top speed week on week. To make it much more of a challenge, the required increase is not simply to add another one kilometre per hour each week. It must be one kilometre per hour the first week, then two the next, and four the next. This is the kind of progress that we have seen in the consumer electronics industry, and it illustrates the challenge that lies ahead.

If he is to achieve his goal, the sprinter will quickly surpass what the human body is capable of, and will have to use ever faster forms of transport to help him. He brings in first a bicycle, then a racing car, a jet aeroplane and finally a rocket. But then he is stuck. The integrated circuit, introduced in 1958, was the equivalent of the bicycle. Very large scale integration (VLSI) is the equivalent of the racing car. Introduced in 1971, VLSI gave integrated circuit manufacturers a way of laying down thousands and then millions of transistors at once. Since then, integrated circuit designers have been tinkering under the hood and making significant and sustained improvements. But soon, we need to bring in the equivalent of the jet plane, and before we know it, the rocket, too. Within about ten years, we will have used all that conventional electronics manufacturing technology has to offer.

And so, to make good the promise of progress, manufacturers will have to utilise new and exciting materials and employ the weird properties of matter that arise only at the tiniest scales – fresh from the minds and chalkboards of theoretical physicists. Fortunately, many of the necessary innovations will begin to emerge from research laboratories quite soon and find their way into the hands of eager consumers. It may not be the end of progress in consumer electronics, but by 2015, it will probably be the end of the silicon chip.

Navigating the Future

Although competition is often fierce in the consumer electronics industry, there is also cooperation. When it comes to charting the way ahead for the technology of integrated circuits, and striving to 'obey' Moore's Law, the industry pools the expertise of chip designers and researchers from across the world. The near-future progress in chip design and manufacture is laid out clearly in the International Technology Road Map for Semiconductors (ITRS), which is produced every two years by a collective of semiconductor industry associations from across the world. The road map identifies various specific targets in chip miniaturisation and manufacturing technology from the present day until 2020, and documents the proposed technologies by which they can be achieved. So, for example, by 2012, according to the ITRS, manufacturers will be making most chips from wafers 45 centimetres in diameter, rather than the current maximum size of 30 centimetres. This will mean that many more chips can be made in one go. Just a simple change like this will reduce the cost of chips – and therefore all kinds of digital devices – dramatically.

The ITRS road map examines several emerging technologies that may help to sustain Moore's Law for many years to come. One such technology that could play a very important role in the development of new generations of processors and memory chips is 'spintronics' – short for 'spin-based electronics'. The name comes from a property of electrons that physicists call 'spin', which is closely related to magnetism. By virtue of its spin, every electron acts as a tiny magnet. In a magnetic field, the spins of electrons either align in the same direction as the field or in the opposite direction. Free electrons – those that are not bound to atoms, such as electrons flowing along wires – will always align in the same direction as a magnetic field. Once outside the influence of the field, the electrons will retain their spin direction until another magnetic field comes along and changes it.

The **move towards multi-core technology** is important in the march towards more powerful processors – in ten years, **100-core processors** may be commonplace

In a conventional circuit, electron spins are aligned at random, and today's chips only make use of electrons' electric charge to represent binary information, not their spin. Inside a spintronics device, magnetic fields align electrons' spins in one of two opposing directions as the electrons speed through. As a result, individual electrons carry either a bit '0' or bit '1', and they can be made to flip between the two states. All modern hard disk drives already make use of spintronics, thanks to the 'giant magnetoresistance effect' that has been responsible for the rapid increase in hard disk capacities since the turn of the century. The next promising application of electron spin in electronics is the development of magnetic random access memory (MRAM). Like flash-based memory, MRAM chips will not lose their information when the power is turned off – so PCs will take virtually no time to boot up. And MRAM will be able to read and write information as fast as, or faster than, conventional RAM chips in today's PCs. Consumer products with MRAM versions should be available as early as 2010.

The holy grail of spintronics research is to make a processor chip that utilises electron spin. Spintronics transistors on such a chip would flip electron spins when carrying out calculations and executing instructions – a much faster and less power-hungry process than turning on and off electric currents as happens in conventional circuits. A basic design for a spintronics transistor was set out as early as 1990, and in 2004, a working prototype was made by Christian Schoenenberger at the University of Basel, Switzerland. This technology is a long way off being incorporated into chips in consumer electronics products. One of the biggest barriers is finding the right materials from which to fabricate a spintronics processor chip. Silicon would be ideal, since then spintronics chips could be made using adaptation of existing chip fabrication technology. In 2007, researchers based at the University of Delaware and a US company called Cambridge Nanotech announced that they had found a way to manipulate electron spins in silicon. But don't expect to buy products with silicon spintronic processors in them just yet: the silicon device had to be cooled to −188° Celsius, and contained no transistors. It was just a proof of the fact that electron spins could be controlled in silicon.

Another way to increase the speed of a chip while reducing its power is to use wireless links within the chip. Having all a chip's interconnections made from tiny silicon or metal wires slows down the chip and wastes energy. In 2007, IBM demonstrated an on-chip device that greatly increases performance of multi-core processors (CPUs with two or more processors working together). The move towards multi-core technology is important in the march towards more powerful processors – in ten years, 100-core processors may be commonplace. However, with current on-chip communications between cores, chips would drastically overheat with more than a few cores chomping away at numbers and communicating with each other constantly. IBM's tiny new device – by far the smallest of its kind – takes binary information from electrical signals and codes them as pulses of light. The light carries the information between the separate cores. According to the researchers, using light instead of electricity increases the speed of communication 100-fold and reduces the power needed by 90%.

Light is not the only medium that can carry information wirelessly within a chip. A team led by Alain Nogaret at the University of Bath is testing an ingenious on-chip wireless communications system using microwaves – the same kind of electromagnetic radiation used by mobile phones. Separate integrated circuits within a single product could also communicate with each other using this system, getting rid of the need for wiring. If the technology turns out to be viable, then it may be used in consumer products by 2015.

How Small Can We Go?

Although the ITRS road map examines some radically new technologies like spintronics and wireless on-chip communications, it focusses mainly on how to continue improving the standard fabrication method for integrated circuits. In other words, it looks at how to achieve the sustained shrinking of transistors on silicon chips fabricated by very large scale integration (VLSI).

The major miniaturisation mileposts marking the way along the road map are called 'nodes', and they are named according to the size of the smallest components etched onto a chip. As I write this book, we are just leaving the '65 nanometre node': between 2003 and 2007, nearly all manufactured CPU chips had transistors 65 nanometres in diameter. A nanometre (nm) is one billionth of a metre, or one millionth of a millimetre, and any process that involves elements less than 100 nanometres across is referred to as 'nanotechnology'. So, the terms 'microprocessor' and 'microelectronics', which have had currency since the 1970s, should now really be replaced with 'nanoelectronics' and 'nanoprocessor'.

In June 2007, the Japanese company Panasonic became the first company to mass-produce integrated circuits from the new node: with transistors just 45 nanometres wide. These chips are designed to code and decode high-definition video signals. They were used in the company's Blu-ray DVD players, a direct replacement of the existing 65-nanometre chip. Panasonic claimed that the new chips were 39% smaller and used 30% less power. In November of the same year, Intel became the first company to mass-produce microprocessors with 45-nanometre transistors. Within several weeks, products were available that contained Intel's chips.

Forty-five nanometres is equivalent to the combined diameter of about 100 silicon atoms. Thousands of these transistors could fit inside a red blood cell. As Intel moved from 65 nanometres to 45 nanometres, they encountered a problem with a layer of insulating material (a substance that does not conduct electricity). The layer was to be just a few atoms thick in places; had they used the traditional material, silicon dioxide, electric current would have leaked through it. Such a leakage would seriously affect the performance of the chip, so Intel replaced the silicon dioxide with a material containing the element hafnium that can do the job without leaking current. They also had to use metal wiring on the chip to connect the integrated transistors together, rather than the 'polysilicon' that has been used for the past twenty years. Other manufacturers have also made similar design changes in order to bring 45-nanometre technology to the fabrication stage. These changes will see them through until at least the next stage of miniaturisation: the 32-nanometre node.

In 2007, several companies, including Intel, IBM and Samsung, demonstrated working prototype chips with 32-nanometre components. According to the road map, products with 32-nanometre transistor technology should become available in consumer products in 2010.

An idea of scale: human hair is typically 100 micrometres (microns) in diameter. This is about one-tenth of a millimetre. A typical bacterium is about one-hundredth as wide, at 1 micron. An HIV virus is about one-tenth as wide again, at 0.1 microns, or 100 nanometres. The width of a DNA double helix is about 2 nanometres. The transistors in today's integrated circuits are 45 nanometres across.

Human hair: 100 micrometres
Typical bacterium: 1 micrometre
Diameter of HIV virus: 100 nanometres (0.1 micrometres)
Width of DNA double helix: 2 nanometres

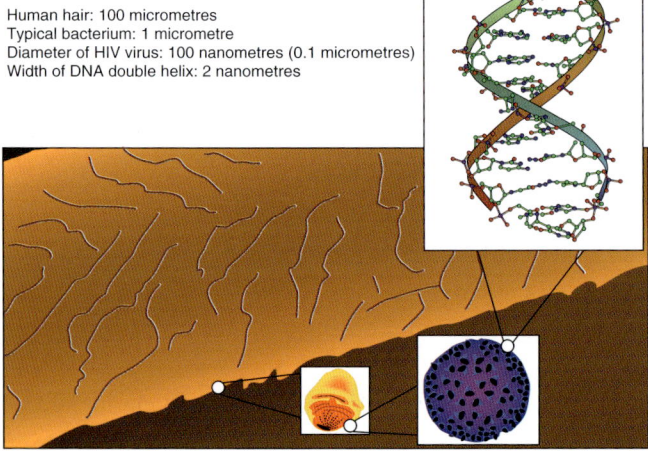

Why So Small?

There are several advantages of having smaller circuit components. Intel's 45-nanometre chips have twice the density of transistors than their 65-nanometre chips have. This means that the same processing power can be achieved with a smaller chip or that a same-size chip could be at least twice as powerful – all for less money. As well as reducing the cost of processing power, smaller transistors require less power to switch. And so, this rapid progress along the miniaturisation road will quickly bring more of the same kind of change as we are already used to: faster, more responsive, more versatile, more powerful and cheaper gadgets.

For nanoelectronics to progress further, many other specific challenges will have to be met and problems overcome besides the leakage of electric current through insulators. One of the most important is overheating. Chips that crunch billions of numbers every second also devour lots of electrical power – all of which eventually ends up as heat. If a chip becomes too hot, it can begin behaving erratically or become damaged beyond repair. This is one reason why the chips inside portable devices are less powerful than those in PCs: if they were as powerful as those in desktop computers and games consoles, they would need fans and bulky heat sinks to keep them cool. Despite the fact that individual transistors will become much more efficient as they shrink, many more of them will be present on a given silicon die, resulting in an overall increase in processing power and therefore heat.

There is another fundamental reason why chips produce so much heat than just the passage of electricity through the circuits. That reason is more subtle than chips 'devouring lots of power' to 'crunch numbers'. In 1961, physicist Rolf Landauer made an important connection between what happens inside a computer's CPU and an area of physics called thermodynamics. When a computer carries out a calculation, information is lost – for example, add two binary numbers and only the new number 'exists' any more. According to the laws of thermodynamics, a loss of information is always accompanied by the production of a corresponding amount of heat. So, even with the most efficient transistors, CPUs of the kind used today will still produce heat – and the more calculations per second, the more heat will be produced. There is a way to avoid this problem; it is called reversible computing. In this context, 'reversible' means that, in principle, computations carried out by the processor could be run in reverse to find out what the inputs were. In other words, no information is ever really lost.

Contrary to what common sense might suggest, this does not mean that the computer's memory would fill up with information unsustainably quickly. By clever programming and a slight slowdown in the rate of computation, reversible computing can be achieved with very little extra storage. Reversible computing is outside mainstream thought – although every computer chip designer is aware of the problem it addresses. And when heat becomes a real barrier to progress when trying to design smaller, more powerful CPUs, hardware and software engineers may have to instigate the necessary changes to make reversible computing a reality. When it comes to it, reversible computing is not all-or-nothing: there are ways to reduce heat that do not involve saving every bit of information. You can think of reversible computing as a way of recycling your information to reduce the power consumption – and when the recycling facility becomes too full too quickly, you can always use the landfill site occasionally. It is about managing information better by appealing to the well-known rules that nature has laid before us.

Too Small to Handle

Beyond the production of excessive heat and the unwanted leakage of electric current, miniaturisation itself presents a significant challenge. When you make structures a few atoms thick, your margin of error is very small indeed. Then there is a more fundamental problem that concerns the wave nature of light. An important part of the current chip fabrication process involves shining light through a mask and onto the silicon wafer. The shadow of the mask defines where the circuit components and connections will be on the surface of the finished chip. But the fact that light behaves as a wave means that the edge of the shadow will always be slightly blurred, however much in focus the light is when it hits the surface. As a result, for a given wavelength of light, there is a limit to how small the smallest component can be.

When **you** make structures a few **atoms thick**, your margin of error is **very small indeed**

Until recently, fabricators used ultraviolet radiation (UV) with a wavelength of 436 or 365 nanometres. In order to make 65-nanometre features, they had to start using lasers that could produce 'deep ultraviolet' radiation, with shorter wavelengths. In theory, the smallest features that can be etched onto a silicon wafer using these wavelengths is 50 nanometres. But several cunning techniques enable manufacturers to push that particular envelope significantly. The most important of these is immersion lithography. In this technology, a liquid fills the gap between the lens from which the UV emerges and the surface of the silicon wafer. The liquid – normally ultra-pure water – slows down the UV, and this has the effect of shortening its wavelength. In 2007, IBM produced features 30 nanometres in diameter using this technique. Eventually, however, 'light' of even shorter wavelengths will probably have to be used: first extreme ultraviolet radiation, then X-rays and even electron beams. At the same time, however, a different approach may make such decreases in wavelength unnecessary. Instead of using a mask and shining light onto a wafer, the components may be physically 'moulded' by using a stamp – just like making indentations in modelling clay with a biscuit cutter – though on a much, much smaller scale, of course. This technology, called nanoimprint lithography, could be good for structures down to about 10 nanometres in diameter.

In immersion lithography, a drop of pure water held between lens and silicon wafer makes it possible to etch smaller details.

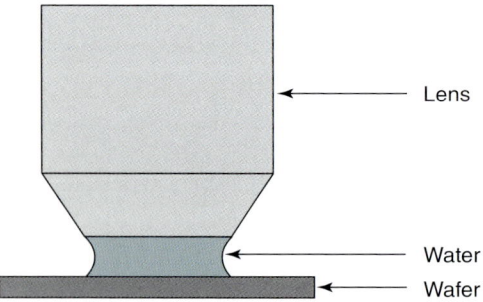

Lens

Water
Wafer

Self-made Chips

As the size of the features on integrated circuits approaches the size of molecules, it becomes more and more of a challenge to fabricate chips using conventional methods. So how will future fabricators make chips with features measuring a few nanometres or in which individual molecules themselves are circuit elements? Devices that can move individual atoms exist, but this is laborious at best; so self-organisation is the only real option when fabricating atom- and molecule-sized features. A self-assembled chip makes itself: the various elements of a chip fall into place by themselves, just as water molecules do when forming an ice crystal. Imagine a set of small blocks that have magnets in certain spots, so that when you throw them into a box, they arrange themselves into a predetermined pattern. Alternatively, chip designers may one day be able to borrow a different technique from nature. Assembler molecules called ribosomes use information from DNA to build proteins that make up our skin and hair, for example, from smaller molecules. Sometime far in the future, tiny nanomachines might one day fabricate our nanoelectronic devices automatically.

In 2007, IBM took an important step towards self-assembling chips, producing what they called an 'air gap' circuit. The air gap is IBM's way around the problem of leaking electric currents in 45-nanometre chips. Instead of laying down new materials, as Intel does in its 45-nanometre chips, IBM's chip is porous, so that there is no material – just a gap. IBM's process uses molecules with a specific shape that build up a repeating pattern when deposited on the surface of the chip. So, for now, this is little more than mimicking nature's way of building porous materials like tooth enamel or seashells. But in the future, more complicated chip features and components may grow themselves – so much more convenient than having to etch details a few nanometres in diameter.

An image of IBM's airgap chip, created in 2007 by self-assembly.

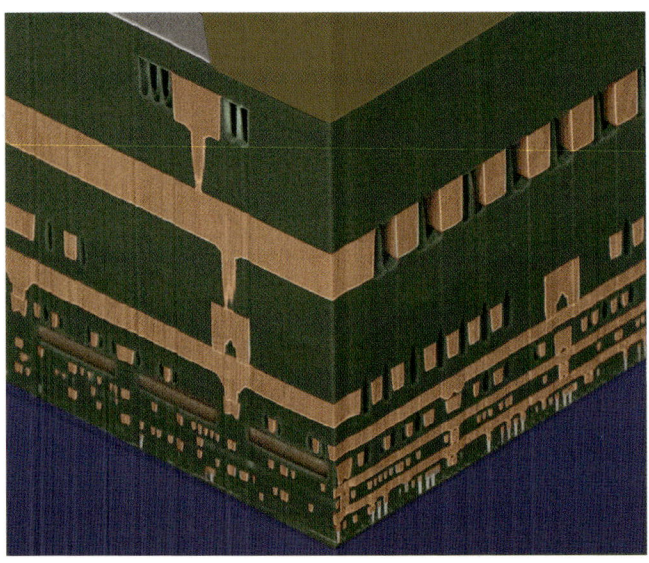

Before the semiconductor industry reaches the 10-nanometre point, it may well have come up against another barrier that could stop Moore's Law in its tracks. When transistors measure 16 nanometres across – perhaps as early as 2015 – a strange effect called quantum tunnelling will prevent them from working at all.

Quantum tunnelling is predicted and explained by quantum physics – the science that describes the often counter-intuitive behaviour of very small objects. The bizarre phenomenon of tunnelling is different from simply leaking currents: it involves electrons literally disappearing from one side of a transistor and appearing at the other. It can only happen over very short distances – and that is why it represents a major roadblock down the line on the ITRS.

Silicon is **cheap and plentiful**, but it is **not** really the ideal material for **making** tiny circuit components

All of the issues described above present major challenges to traditional VLSI technology, even with the innovative approaches also described above. However, many of the problems lie not in the manufacturing process, but in the materials being used – and silicon in particular. Silicon is cheap and plentiful, but it is not really the ideal material for making tiny circuit components. It takes a good deal of energy to force electrons through it – this is why silicon chips are power hungry and one reason why they run hot. And those electrons only travel relatively slowly through silicon, so microprocessors based on silicon run more slowly than chip designers would like.

Goodbye Silicon

Fortunately, there are other materials that can be used to make tiny transistors, yet which do not have the problems associated with silicon. Eventually, the term 'silicon chip' will probably fall out of fashion. One class of materials long known to have similar electronic properties to silicon, but without the same problems at small scales, is the 'compound semiconductors'. While silicon is a semiconductor in its pure, elemental state – and is used in that way in silicon chips – compound semiconductors are composed of two or more elements, either mixed together or chemically combined. There is a huge range of compound semiconductors, and they do not suffer from the quantum tunnelling problem at very small scales. One of these materials in particular has already found wide application in a number of semiconductor devices. Gallium arsenide – a simple alloy (mixture) of the elements gallium and arsenic – has been used to make chips for military and space applications, and even mobile phones, since the 1980s and in LEDs (light-emitting diodes) since the 1960s, for example. However, gallium arsenide has its own drawbacks – including the fact that it is much more difficult and more expensive to process than silicon. Nevertheless, now that miniaturisation based on silicon is threatening to slow down or stop, researchers are looking at ways to make these compound semiconductors viable alternatives to silicon in consumer products. In 2006, Intel produced a transistor made using the compound semi-conductor indium antimonide. It was nearly twice as fast as their silicon-based transistors and used one-tenth as much power.

While compound semiconductors are one hot topic, many researchers are focussing their attention on a set of materials discovered and investigated more recently: fullerenes, carbon nanotubes and graphene. These are all forms of pure carbon in which the carbon atoms join together in interconnected hexagonal, pentagonal and heptagonal rings. The result is single-atom-thick sheets of carbon in spheres, cylinder and planar formations. They are very strong, fairly easy to make, and they have an interesting range of electrical properties. In fifteen years, when the technology has matured, they might just keep Moore's Law on track when silicon has given up the ghost. We want enormously powerful, portable devices that can be programmed to have intelligence and constant access to information – and yet still draw little electrical power but stay cool. Look no further than carbon; it could be the new silicon.

Fullerenes

Before the discovery of fullerenes in 1986, only two forms of pure carbon were known: diamond and graphite. The scientists that made the discovery – a team led by Harold Kroto at the University of Sussex – were following up an observation made using radio telescopes. Astronomers had detected signals coming from interstellar space that indicated the existence of molecules of pure carbon of a type unknown on Earth.

The team decided to vaporise graphite to produce individual carbon atoms and cool them to form clusters. They found that clusters of 60 carbon atoms were very stable, and they quickly worked out that the clusters were spherical closed molecules just less than a nanometre in diameter. It became apparent that the carbon atoms in these molecules were joining together to form interlocking hexagons and pentagons. Kroto and the team named the newly discovered molecule buckminsterfullerene, after the British architect Richard Buckminster Fuller, famous for designing geodesic domes made from hexagons and pentagons. New fullerenes soon emerged: smaller spheres, larger spheres, rugby ball-shaped molecules, balls within balls arranged like Russian dolls. And soon there emerged other similar forms of carbon that are not fullerenes but have similar structures. Most important in the new world of carbon are sheets of carbon atoms in hexagonal arrangements either flat (graphene, 2004) or rolled into tubes (carbon nanotubes, 1991).

Kroto and his team were awarded the 1996 Nobel Prize for chemistry, and fullerenes have become much more than a scientific curiosity. They have a wide range of potential applications, thanks to their exceptional strength, their various electrical properties and the fact that they can now be made cheaply.

Shaped like a football, buckminsterfullerene is a molecule of 60 carbon atoms joined in hexagons and pentagons, which spawned a huge new crop of carbon-based molecules: the fullerenes.

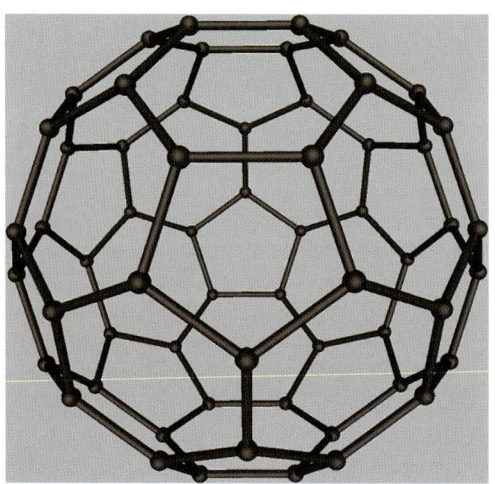

One form of pure carbon that may prove to be important in nanoelectronics is graphene – first produced in the laboratory as recently as 2004. In graphene, the hexagonal carbon rings join together forming a flat sheet. Graphite, familiar as the main constituent of pencil 'lead', is effectively a stack of the single-atom-thick graphene sheets. Graphene was long deemed impossible to produce, because it was assumed that it would be too fragile or would simply curl up to form tubes. In 2007, a team at the University of Manchester led by Professor Andre Geim was able to make the world's first working graphene transistor, by 'carving' a sheet of the material using a beam of electrons. This was the same team that made the first samples of this strange material in 2004. Geim's team managed to carve a

This photograph shows the graphene transistor made by Andre Geim and his team in 2007. The central 'island' of graphene is less than 100 nanometres wide, and the contacts between it and the larger connections (the transistor's source and drain) are just a few nanometres wide. These contacts allow only one electron at a time onto the island. The two 'large' gold pieces are metal connections to a test circuit.

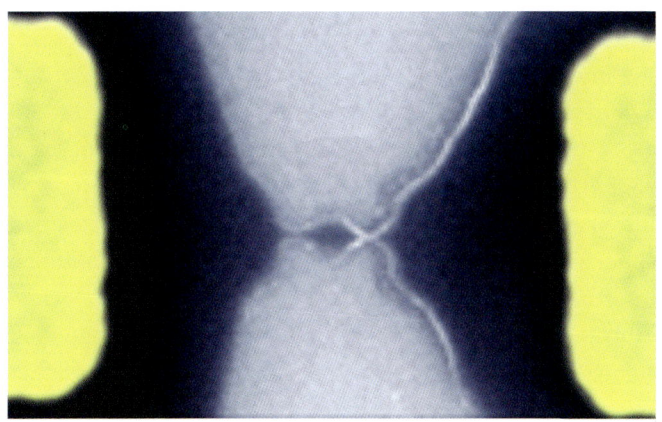

tiny island of graphene in contact with a small sheet of the material on either side. The whole single-atom-thick transistor was just 50 atoms wide – and in the future, transistors made in this way could be much smaller, and they would perform well down to the smallest scales. In fact, it may well be possible to produce a transistor that consists of a single hexagonal ring of carbon atoms. It is quite possible that a complete chip could be carved from a single sheet of graphene. Such a chip would produce much less heat than a silicon-based transistor, and would be significantly faster. However, even optimistic estimates put this development fifteen years into the future.

Black Gold

The most exciting, and certainly the most versatile, in the crop of new carbon nanomaterials are carbon nanotubes. These minuscule tubes of soot hold the promise of replacing silicon as the main material for making processors, memory chips, and for use in display screens, batteries and solar cells. Carbon nanotubes are hollow cylinders that consist of hexagonal rings, and you can picture them as chicken wire rolled up into tubes – although, of course, about a hundred-millionth the size. Carbon nanotubes can be as long as a few millimetres but as narrow as 2 nanometres. Each one is a single molecule, and is capped at each end by a hemisphere. Twisted together like twine, carbon nanotubes can be made into rope that is as strong as a thick steel cable. A 'forest' of carbon nanotubes grown on an ordinary silicon chip can act as an efficient cooling device for conventional electronics, thanks to its extraordinary ability to conduct heat. But it is the electrical properties of individual nanotubes that are of great interest in the future of electronics. Depending upon their exact shape and size, carbon nanotubes can be excellent conductors of electricity or reliable semiconductors. Several researchers have made transistors using single carbon nanotubes, and these minuscule devices have outperformed their larger silicon counterparts.

Carbon nanotubes are hollow **cylinders** that consist of hexagonal rings, and you can picture them as **chicken wire rolled up into tubes**

Nanotube Transistor

The most simple arrangement for a carbon nanotube transistor, developed by IBM, comprises a single semiconducting carbon nanotube laid across two gold electrodes – the 'source' and the 'drain'. The nanotube is normally non-conducting, so that no current passes from source to drain; the transistor is 'off'. It is in contact with a layer of silicon dioxide, which insulates it from the (conducting) layer of silicon below. A voltage applied to the silicon layer creates an electric field that affects the conductivity of the nanotube, allowing current to pass and turning the transistor 'on'.

However ideal nanotube transistors might be in theory, they currently have to be made to order individually, and it can take several days to produce a single one. If chips based on carbon nanotubes are to become the next big thing in computer processors and memory chips, some way will have to be found of making millions of transistors at a time.

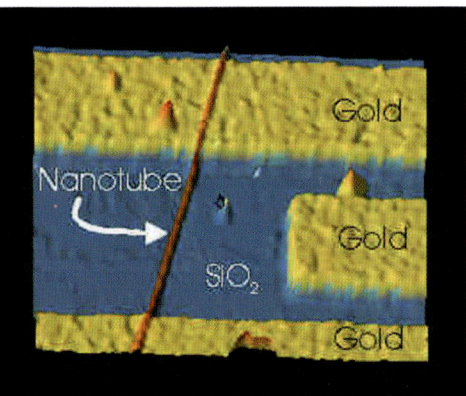

Making carbon nanotubes is now relatively easy – millions of them can be made in a simple lab set-up. And it will become easier and cheaper as demand for these tiny cylinders grows. There is a problem, however: when carbon nanotubes are made *en masse*, the result is always a tangled, random mixture of the conducting and semiconducting forms. As long ago as 2001, researchers at IBM found a way to destroy the conducting ones, leaving only the semiconducting ones behind – those that are most important in producing tiny transistors. This is less than ideal, of course, since it represents a real waste of space; also, you may be left with the desired type of nanotubes, but they are still in a tangled mess. Self-assembly will probably be a way forward, but that seems a long way off, since at present, no one has any idea how to achieve it. In 2003, a team at a Californian company called Nanomix, together with researchers at the Naval Research Laboratory in Washington, DC, found a way to produce a carbon nanotube sheet from only semiconducting carbon nanotubes. Laid on a silicon substrate, the sheet behaved as a transistor. Nanomix has gone on to produce medical sensors – for detecting blood glucose, viruses or specific sections of DNA – by attaching specific chemicals to the nanotube sheet. When the 'target' molecule attaches to the chemicals, the electrical properties of the nanotube sheet change, allowing a meter to display the presence of the target.

For some applications, it is possible to utilise the jumble of tubes as they are – without selectively destroying either the conducting or the semiconducting ones. Laying carbon nanotubes flat between electrodes produces a highly conductive sheet of tiny wires that provides thousands of different routes for electrons to pass from one electrode to the other. This sheet can be made transparent and flexible, which can be useful as a surface conducting sheet in solar panels or touch screen displays. Alternatively, millions of the tubes can be suspended in a liquid and then sprayed onto a surface. This technique promises cheap and flexible display screens, electronic paper, and solar cells. The spray-on carbon nanotubes can be applied to a printing block, so that they can be printed onto any surface.

Flexible, printed circuits would be useful in making wearable computers, or in embedding electronic devices in fabric or paper.

One application of carbon nanotubes that is about to find its way into consumer electronics products is nano-RAM, or NRAM. Developed by a company called Nantero, based in Boston, Massachusetts, NRAM chips each holds 10 gigabytes and can be fabricated with only a slight extension of existing silicon chip fabrication techniques. NRAM chips allow for very fast storage and retrieval of information, need very little power, and unlike conventional RAM, do not lose the information they hold when the power is off. Like MRAM, NRAM will allow PCs to boot up immediately; and it could compete with existing solid-state memory in replacing hard disks in laptops. This would certainly improve battery life. NRAM could also provide fast, high-volume storage in smartphones and portable media players.

While semiconducting carbon nanotubes could be the new silicon, eventually the new silicon may not be a semiconductor at all. In the move towards nanoscale computing, researchers have to consider radical solutions if they are to avoid a head-on collision between Moore's Law and the laws of nature. So, rather than simply trying to get carbon nanotubes to fit into the shoes of silicon thanks to their ability to be semiconductors, why not abandon semiconductors altogether? You can process digital information without semiconductors. In 2005, Hewlett-Packard suggested doing away with semiconductors when they announced a technology they call a crossbar latch. This extremely tiny nanodevice consists of very thin platinum wires (nanowires) lying across each other to form junctions. A single 'switchable' molecule is trapped between the two wires at each junction. As a result, the molecule can store a binary '0' or '1' simply by accepting or losing a single electron; and in the same way, it can take part in 'logical operations' – manipulating binary digits to carry out instructions and calculations. There are many technical issues to sort out before this nascent technology can compete with the established approach. For example, the molecule only lasts around a hundred 'switches'. But if and when these issues are sorted, the crossbar latch could re-revolutionise the digital revolution. The researchers at Hewlett-Packard suggest that the technology may be commercially viable – for memory chips at least – as soon as 2012.

Memories Are Made Of . . .

While we are on the subject of memory, there are several other technologies that may one day give us practically unlimited storage in small, low-power devices. The first, holographic storage, involves writing information as holograms – three-dimensional images created in a material by the 'interference' of laser beams. To achieve this, binary information is represented in a chequerboard pattern on a device called a spatial light modulator – something like an LCD with a million light and dark pixels. A single laser beam splits in two: one passes through the spatial light modulator and the other remains unchanged as a 'reference beam'. The

The crossbar latch, a molecular switch that could one day replace transistors as the basis of processors and memory chips.

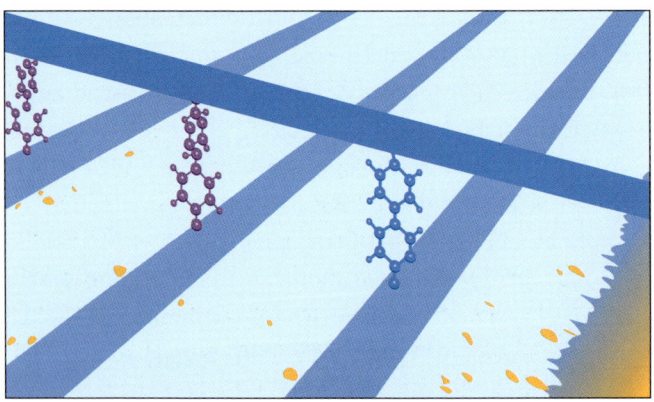

The idea of a crystal storage device the **size and shape of a sugar cube** that is capable of storing **terabytes** of information (thousands of gigabytes) has been anticipated for **at least thirty years**, and will probably become reality in the next **ten years**

holographic image of the chequerboard pattern forms where the two beams meet. In a recent, but existing, form of this technology, the hologram is recorded on a disc containing a photosensitive gel. The disc, not much bigger than a DVD, can hold 300 GB of information – compared to just 4.8 GB for a standard DVD disc and 50 GB for a Blu-ray disc. For now, this technology is used as an expensive, professional archiving device; but it will almost certainly go the way of many technologies that start off expensive: it will probably become cheap enough to find its way into consumer devices before too long. And other forms of holographic storage will probably follow: the idea of a crystal storage device the size and shape of a sugar cube that is capable of storing terabytes of information (thousands of gigabytes) has been anticipated for at least thirty years, and will probably become reality in the next ten years.

The second potentially revolutionary mass storage device, which could replace hard disk drives and perhaps even RAM, is IBM's Millipede. This takes the form of a chip rather than a disc, and represents binary digits with tiny moving parts rather than using magnetism or a holographic image. On the surface of the chip is a very thin film. To write a '1', a tiny heated probe presses the thin film down, making a tiny pit. To read a '1', the probe passes into the pit it has made, and its tip cools down slightly. The read/write head consists of an array of perhaps 32×32 probes, so 1,024 bits (1 kB) can be written at any given instant. Amazingly, this system is robust enough that production models may soon be

In a holographic storage device, binary information is coded in one laser beam, which produces an interference pattern when it meets another laser beam on a light-sensitive disk. A gadget using this technology has the potential to store immense amounts of data in a small space.

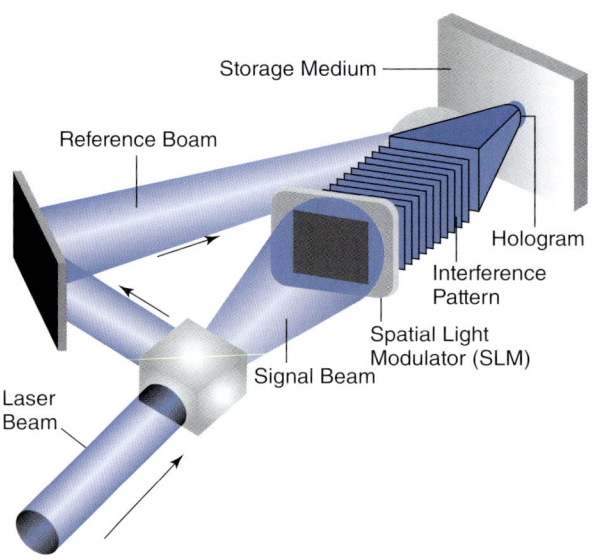

Storage Medium

Reference Boam

Hologram

Interference Pattern

Spatial Light Modulator (SLM)

Signal Beam

Laser Beam

available, giving storage densities far in excess of either hard disks or flash memory chips, and with read/write times as fast as RAM. Millipede promises storage densities of about 1 gigabit per square millimetre (1,000 gigabits per square inch) – more than four times the corresponding value for the latest hard disk drives. In addition, Millipede chips should be capable of reading and writing information at speeds much faster than hard disk drives, flash memory and conventional RAM.

IBM's Millipede storage chip stores binary '1's as tiny indentations in a polymer surface. The device can store and access huge amounts of information extremely rapidly.

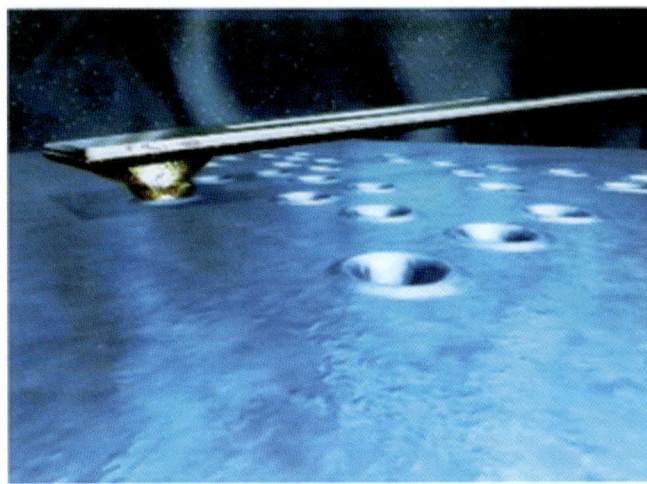

A little further into the future, we may have devices in which each binary digit is stored in a single atom. In 2007, IBM researchers demonstrated the potential of this technology, by announcing that they had successfully measured the magnetic state of individual iron atoms with a tiny probe. If this technology could one day become part of a storage device, then a product the size of an mp3 player would be able to store 30,000 full length, high-definition feature films. It is already possible to map out, probe and even move individual atoms – using a device called scanning tunnelling microscope or a similar device called an atomic force microscope. But clearly this is no way to make a mass storage device: it would take an unfeasibly long time. Again, when things get down to the scale of molecules and atoms, self-assembly is the way forward, but no one knows how that might work – yet.

In 2007, researchers at the University of Pennsylvania published their work on another nanoscale, high-volume information storage device. They managed to make use of self-assembly to create 'nanowires' of a material called germanium antimony telluride using a vapour of germanium, antimony and tellurium. This material acts as a catalyst that helped these elements form spontaneously into wires just 30 nanometres in diameter and 10,000 nanometres (0.01 millimetre) in length. The wires' atoms change from an organised crystal structure to a disorganised 'amorphous' structure incredibly quickly when they are heated. As a two-state system, this system can store binary information. If produced, it would be capable of retaining data for tens of thousands of years and of reading and writing a thousand times faster than existing portable memory devices, while using less power.

New Technologies on Display

The future is not only about new improved CPUs and memory. Incredible advances are also being made in display technology, too. It seems like only yesterday that

We all **love** **wide, thin, flat screens** capable of displaying high-definition TV; this has been the **main draw** away from **cathode ray** tube televisions and towards **LCD** **and plasma displays**

LCD and plasma television displays became popular, and they are only just becoming affordable. But the dominance of these two technologies is under threat, from alternative technologies that will be cheaper to make (and eventually cheaper to buy), lighter and less power-hungry. LCDs and plasma screens will soon feel old-fashioned. And the LCD computer monitor? Its days are numbered, too. Prepare to be dazzled – and prepare to see low-power, high-resolution images on a host of surfaces. So what are these new technologies, and what else will be on offer to us, besides screens for our TVs and PCs, to display our digital information?

We all love wide, thin, flat screens capable of displaying high-definition TV; this has been the main draw away from cathode ray tube televisions (CRTs) and towards LCD and plasma displays. But our current favourite displays have a number of disadvantages, too. Firstly, both use a lot more power than CRTs – although this is because we like our thin screens to be considerably larger than our bulky CRTs ever were. But there are other reasons why these technologies are inefficient. An LCD's backlight is always on, even when a dark picture is being displayed: the pixels just block most of the light (this also means that 'blacks' do not look truly black). A plasma simply uses more power anyway – just because of the way it produces its picture. Secondly – with LCDs at least – fast-moving objects in scenes can look blurred. And finally, there is the fact that these televisions are considerably more expensive than CRTs. The new breeds of televisions and other display devices address most or all of these concerns. But which one to choose? The array of new and new-ish technologies is quite staggering. There is SED, FED, LCoS, DLP, laser TV, OLED, IMoD and low-power LED backlighting for LCDs.

Inside a field emission display (FED), an array of fine carbon nanotube spikes sits just behind a phosphor-coated screen. When a negative voltage is applied to the spikes, they produce an intense electric field at their tips, which drives electrons off towards the screen. This technology was developed a few years ago, but instead of carbon nanotubes, it used an array of metal spikes that was prohibitively expensive to produce. Carbon nanotubes have given FED technology a new life. Surface-conduction electron-emitter display (SED) technology works in a very similar way, but the electrons are produced by a flat layer of carbon with tiny gaps in it. Both types of display produce a picture with the clarity of a CRT television but in a large, thin and more energy-efficient package.

SEDs and FEDs are both similar to LCDs and plasma displays in that they produce a picture directly on the screen. Most other new display technologies involve projection. The image is produced inside the television, and then projected onto the screen. Projection technologies include liquid crystal on silicon (LCoS) and digital light processing (DLP). LCoS contains a tiny liquid crystal display on a reflective silicon surface. The whole thing is a chip that can fit in your hand. DLP is a tried and tested technology that has been used in projectors for about ten years; again, it is based on a chip, but this time, millions of tiny tilting mirrors do the job of producing a picture. In both of these cases, the light source is normally a bright white bulb whose light is split into red, green and blue. But a set of three lasers (red, green and blue) can do the same job using less power and with greater clarity and colour precision. Laser TV has had a difficult beginning, but competitively priced laser TV sets should be available by 2009. You might think that to project an image, the displays have to be deep and therefore bulky, but in fact, with clever optics – and especially if they use lasers to produce the image – these televisions can be as slim as other flat panel displays.

One of the problems with incorporating video and photo capabilities into handheld devices is that the size of display makes it difficult to watch comfortably. Now, a few companies, including Texas Instruments and Microvision, have developed tiny DLP chips that they plan to incorporate into their mobile phones, so that a phone can be used to project photos, videos and other information onto a nearby surface. These chips will also probably turn up in personal media players, cameras and camcorders. And in a few years, when this technology is cheap, these tiny devices could be positioned around the room in relevant spots, so that you would have displays wherever you might need them.

It will be interesting to see which of these display technologies wins out in the marketplace over the next few years. However, in the longer term, another technology will probably be more appealing and eventually cheaper: the organic light-emitting diode (OLED) display. This has the best combination of features: it

It will not be too far off that you can project the photos and videos you have taken with your smartphone – directly from your smartphone.

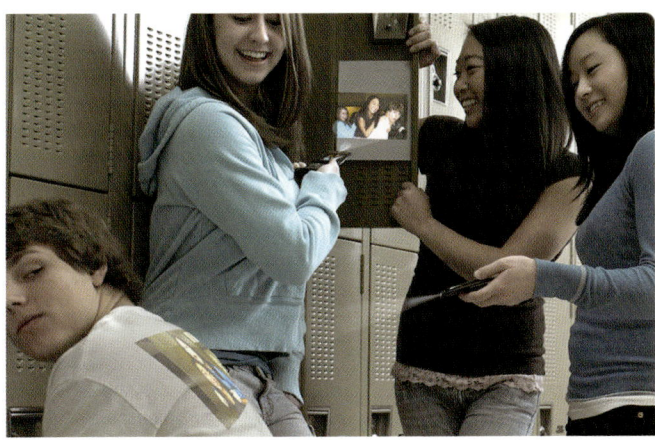

is ultra-thin, draws little power, produces a crisp picture with faithful colour reproduction and very black blacks, and has a wide viewing angle. However, until recently, the materials that produce the light – the organic diodes – have suffered from relatively short lifetimes. In 2007, Sony became the first company to produce OLED TVs for the consumer market. At first they only released an 11-inch model, and only in Japan at first. This was only for the serious early adopter, since even this small model cost about £1,200. Nevertheless, at just 3 millimetres thin, Sony's OLED TVs were stylish and very desirable, and they sold incredibly quickly. Sony, Samsung and Toshiba have produced larger prototype OLED TVs. There are several different types of OLED displays, including those that are almost completely transparent, so that they can be seen from either side, and even flexible, foldable models.

OLED displays have a wide range of other applications beyond television sets. Small displays are already used in a number of consumer products, including some mobile phones, cameras and personal media players. Transparent OLEDs could be used in 'heads-up' displays in cars or in goggles that allow the wearer to see information while also looking at the world around. An Edinburgh, Scotland-based company called MicroEmissive Displays is preparing to produce goggles with OLED displays that can plug into a mobile device, so that you can watch films in large-screen format right in front of your eyes. Such devices are already available with small LCD screens, but the quality has never been quite satisfactory, and the goggles have always been a bit bulky and power-hungry. OLED is soon set to solve all these problems, and to produce brighter, clearer images. Non-display OLEDs can be made into large sheets that can replace fluorescent lights with a range of ambient light options. Perhaps most exciting is the opportunity offered by OLED displays for lightweight, foldable displays that could take the place of newspapers, magazines and books. They could be pocket-sized, fold out to be as large as a magazine, but would load customised news and features. If the electronics was powerful enough, they could display moving images – in case you have seen or read the

Harry Potter series, think of the *Daily Prophet*. Instead of throwing your newspaper away, leaving it on the train or, at best, recycling it, you will simply update it the next day.

It has been a long time coming, but flexible, reusable paper that is easy on the eye really is on the horizon.

How electronic ink works.

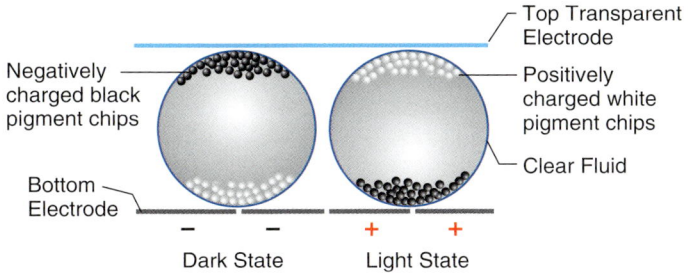

Top Transparent Electrode

Negatively charged black pigment chips

Positively charged white pigment chips

Clear Fluid

Bottom Electrode

− − + +

Dark State   Light State

Mobile Library

There are other technologies besides OLEDs that promise flexibility. In particular, there are many separate types of 'electronic paper' or 'electronic ink' in development – including some that have already found their way into consumer products. And in 2007, the British company Plastic Logic began construction of the world's first large fabrication plant for organic electronics, in Dresden, Germany. Their products will have a range of applications, and they will be concentrating on thin and flexible electronic paper displays.

Electronic paper is already found in many consumer products: there are several stand-alone electronic book readers on the market, for example. They can download and store dozens or even hundreds of books, and display them using electronic ink. The display looks good: it is just as visible in bright sunlight as is paper, since there is no backlight nor does the screen emit any light. This gives electronic paper the edge over OLED displays for electronic books: the fact that it can still display an image without any power is an important property. It also reduces eye strain that can arise from glaring at an illuminated surface.

To make electronic paper, tiny spherical globules are suspended in a thin plastic sheet. The globules are coated with white, positively charged pigment on one side and black negatively charged pigment on the other. Electrodes on the top and at the bottom produce an electric field in one direction or the other to turn the globules when necessary; the electrodes are arranged as a matrix, so that pixels can be addressed individually. The top electrode is invisible, so that light can strike the electronic paper, bouncing off the 'white' globules and being absorbed by the black ones – exactly what happens with black ink and white paper. Electronic paper only uses power to turn the globules, so it is very energy-efficient.

Several companies are working on colour versions based on the same technology, with Fujitsu and LG Philips leading the way. At the 2008 Consumer Electronics Show in Las Vegas, LG Philips showed off a prototype A4 sheet of flexible electronic paper that can display 16.7 million colours at the same resolution as a laptop screen – but is only 0.3 millimetre thick.

For now, the displays in e-book readers are far too small, and they do not have the authentic feel of a real book. The problem seems to be that the supporting electronics takes up more space than the display. But in the future, the electronics will be smaller and much more hidden, so that the electronic paper – once unfolded or unrolled – feels like a sheet or a few sheets of paper. E-ink has also been incorporated into watch displays, personal media players and mobile phones, as well as, inevitably, advertising boards.

In the **future**, the electronics will be smaller and much more hidden, so that the **electronic paper** – **once unfolded or unrolled** – feels like a **sheet** or a few sheets of **paper**

Most flexible displays – including flexible OLED screens and electronic paper – have organic polymers as the basis of their electronic connections and the simple supporting circuit elements, printed or sprayed onto a flexible plastic substrate. Organic electronics – the use of plastic electronic components and connections – is a growing field. It uses materials based on carbon compounds, notably plastics and other polymers. We think of plastics as very good insulators, and most are. That is why power cables and mains plugs are sleeved and insulated with plastics, for example. In the 1960s and 1970s, researchers managed to produce plastics that are good conductors, and since then, a range of semiconducting polymers. Interestingly, the class of polymers most important to organic electronics are the melanins – the same type of molecules that act as pigment in humans and other animals. The melanins and other materials used in organic electronics are cheap to produce, and plastic electronic circuits can be fabricated cheaply in much less strict conditions than those needed to make silicon-based integrated circuits.

Organic electronics has advanced to a state where the components necessary to make a CPU or other useful circuits can be made quite small and integrated onto a single sheet. However, the components are nowhere near as small nor as powerful as good old silicon. And so, appealing though the idea of low-cost plastic electronics may be – and however great their potential applications and advantages might be – plastic circuits are probably not going to be as powerful as silicon-based ones already are. That creates a dilemma: for flexible displays that do more than display text and pictures, you need some processing power – but silicon, which can deliver that power, is not flexible. In 2005, John Rogers at the University of Illinois at Urbana-Champaign came up with a solution. He found a way to make stretchable, flexible silicon – by cutting it into 100-nanometre ribbons and sticking them, crumpled up, onto a flexible rubber substrate. Rogers fabricated several key electronic components using his stretchable silicon technology, and they all performed as well as they would have done in a conventional circuit, despite the fact that he flexed them to and fro. If Rogers' idea catches on, we really can look forward to moving images and web links being included to that newspaper delivered on tomorrow's flexible OLED displays or electronic paper.

The Third Dimension

Another important display technology – that we have been anticipating since even before watching R2D2's message from Princess Leia in *Star Wars* – is 3D-TV. We don't want to have to wear red and blue goggles nor wear a bulky virtual reality headset to view 3D content. Display systems that can produce the 3D effect without the need for headgear are said to be autostereoscopic. Such displays would be a particular boon to gamers, and if and when they become widely available, feature films could be routinely shot in 3D. It always seems a shame that computers and games consoles can carry out powerful 3D rendering, representing three-dimensional virtual worlds inside their processors only to display a two-dimensional version of it on a monitor or TV. Imagine wandering around Second Life or some other online virtual world in glorious 3D and receiving high-definition 3D television programmes. This is not going to happen just yet, but it is such an iconic feature of visions of the future that if it is possible, it will happen.

There are some emerging technologies that can present us with three-dimensional images without the strange headgear – but they are far from perfect, and they will not be making an appearance in your living room just yet. Several companies have developed displays that direct the left and right images at different angles into the viewer's two eyes, thanks to a bumpy, transparent 'lenticular' grating on top of the screen itself. This is how those '3D' picture cards work – the ones sometimes given away in cereal packets – only better. Japanese mobile phone company NTT-Docomo is developing a version of this technology for portable devices. A camera embedded in the device monitors the user's position relative to the screen, and tailors the position of the two images accordingly.

There are many other approaches to 3D displays, with varying success and with varying potential to be developed and incorporated into future gadgets. Perhaps

One **emerging** technology that could be **used as a personal television screen**, in 2D or **3D**, or to provide an 'augmented **reality device**', is the 'virtual **retinal** display'

the most interesting 3D display technology is a 'plasma array' developed by Japan's National Institute of Advanced Industrial Science and Technology. The device creates dots of glowing plasma in mid air by focussing powerful pulsed infrared laser beams. This system can produce dotty images within a volume of a cubic metre, but it could be scaled up. It is doubtful whether the display's very poor resolution can be improved or whether a full colour version could be made – but I would never rule it out.

One emerging technology that could be used as a personal television screen, in 2D or 3D, or to provide an 'augmented reality device', is the 'virtual retinal display'. In this technology, a weak laser scans an image directly onto the retina; the back of your eye becomes the television screen itself. This approach was originally developed for military 'heads up' displays. Research suggests it could even work at a distance – say, from a mobile phone. By actively following the movement of our eyeballs and compensating for our motion, the virtual display could remain steady.

Power to the People

Another vital element of consumer electronics gadgets – besides processors, memory and display technology – is electrical power. Fortunately, the technology

In a virtual retinal display, LED lasers scan across the back of your eye, producing an image directly on your retina.

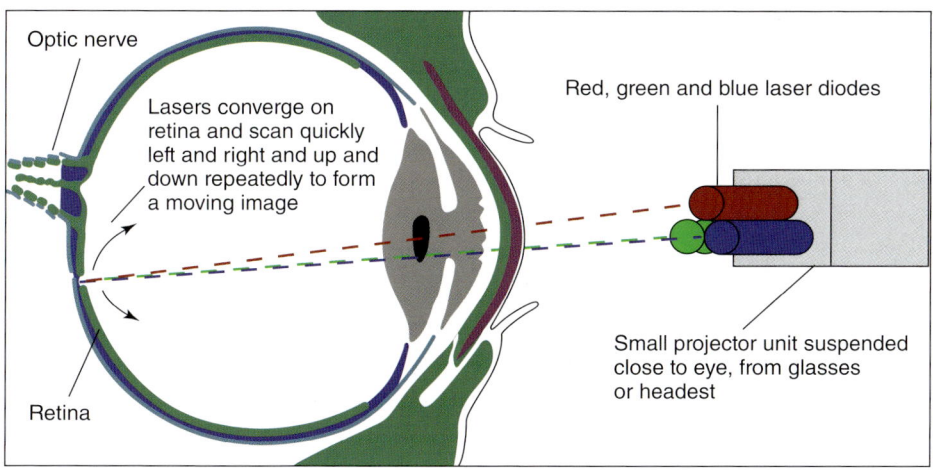

Optic nerve

Lasers converge on retina and scan quickly left and right and up and down repeatedly to form a moving image

Red, green and blue laser diodes

Small projector unit suspended close to eye, from glasses or headest

Retina

that powers our gadgets is also undergoing rapid development. Soon, you will not have to worry about plugging in your devices so regularly, and you may even be able to get rid of your power cables altogether.

Have you ever dreamed of ridding your living room of mains cables and plugs? If you can send information wirelessly, why not send power through the air too? The idea is not new – it dates back to at least an idea by the Croatian-born genius of electromagnetism, Nikola Tesla. But now, researchers at the Massachusetts Institute of Technology have achieved it – in the laboratory at least – using a system they call 'WiTricity'. In 2007, the researchers managed to transmit 60 watts of electrical power across 2 metres – from one coil to another – to power a light bulb. The system uses well-known principles of physics: one coil produces an oscillating electromagnetic field that 'induces' a current in the other one. But it has never been attempted in this way before. In case you are wondering whether there is a threat to human health in a room with oscillating electromagnetic fields flying around transmitting a significant amount of power through the air in this way, you do not have to worry. The WiTricity system produces electromagnetic fields that oscillate a few million times every second (MHz) – frequencies that can have no effect on the human body, even at tens of watts of power.

In another system designed to rid us of wires – and in this case, to rid of us of battery chargers, too – Takao Someya at the University of Tokyo has designed a system of flexible interconnecting 'tiles'. Each tile carries small coils of copper wire, along with some organic electronic components. They produce oscillating electromagnetic fields around the tile; the idea is that you could charge portable gadgets simply by placing them down on the tiles. The special tiles could be incorporated into floors, walls and even furniture at home and at work. To make this work, each gadget would have to have a simple coil inside to receive the induced power. Some gadgets already have these coils, since a system like Someya's already exists: British company SplashPower makes power 'pads' for house and car that work in the same way.

One piece of good news is that our gadgets will need less power – thanks to displays that require no backlight, lower-power processors and the increasing use of chip-based memory solutions rather than hard disks

It is not only the way we charge our portable devices that is changing. The way our devices' batteries hold on to the charge is also changing. Rechargeable

One of Takao Someya's interconnecting tiles that can transmit power wirelessly to nearby devices.

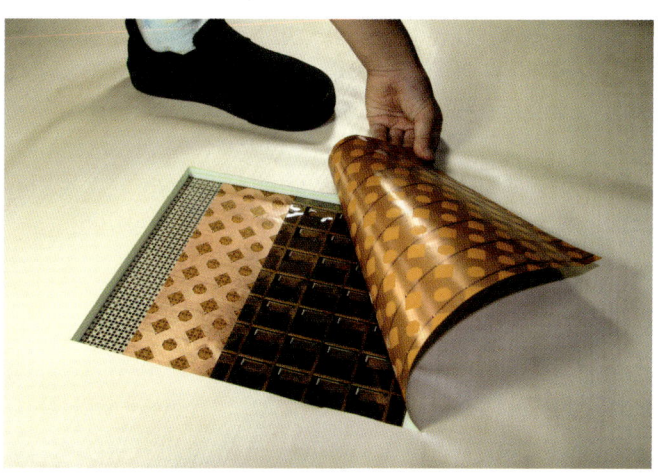

batteries have been around since the middle of the 19th century. The lead-acid battery used in cars is no good for portable consumer electronics: it is wet and uses corrosive sulphuric acid. The nickel cadmium (NiCd) battery, invented in 1899, was the first type of rechargeable battery used in portable digital gadgets from the 1970s. It has been largely superseded by nickel metal hydride (NiMH) and lithium ion (Li-ion) batteries, due to their larger capacity and power capabilities. However, as we all know, these batteries run out of power all too easily and quickly – and often at the wrong moment. So what can new battery technology offer us? How will we power our increasingly portable future?

One piece of good news is that our gadgets will need less power – thanks to displays that require no backlight, lower-power processors and the increasing use of chip-based memory solutions rather than hard disks. However, we will be relying increasingly on wireless communication, which is a major drain on a battery's charge. So, we will still need a reliable, convenient source of power. Carbon nanotubes are playing an important part here, too. In 2007, researchers at the Rensselaer Polytechnic in New York State demonstrated a rechargeable battery that has the look and feel of blackened paper. The 'nanocomposite battery' is mostly cellulose – as is ordinary paper – but it is infused throughout with conductive carbon nanotubes, along with a dry electrolyte. The finished product is cheap to produce, and can be cut to any shape and size, making it an ideal solution to power the range of low-power gadgets in tomorrow's world.

Perhaps an even more attractive portable power supply for the future was invented by Joel Schindall at the Massachusetts Institute of Technology. Schindall invented a type of supercapacitor – a device that stores electrical energy as an electric field rather than through a chemical reaction as in a battery. A capacitor is basically comprised of two metal plates held a short distance apart; the energy-storing electric field exists between the plates. The capacitor can discharge quickly or slowly, depending upon the circuit to which it is connected. The amount of energy a capacitor can store is limited by the surface area of the plates. By coating the

The flexible, paper-thin battery produced at Rensselaer Polytechnic uses carbon nanotubes to store and conduct electricity. Chemicals in sweat or blood could provide the energy for the battery.

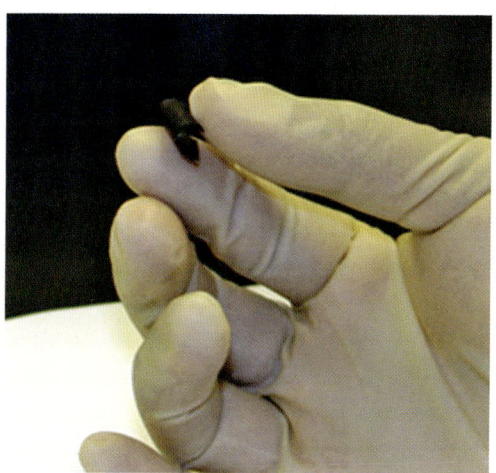

plates with thin carbon nanotubes, Schindall was able to increase the plates' surface area by a factor of many thousands, making it possible to store as much energy as a rechargeable battery. The carbon nanotube supercapacitor can be charged in seconds rather than hours, and will be good for hundreds of thousands of charge-discharge cycles, compared to no more than a thousand for today's rechargeable batteries.

Another technology that may be powering our gadgets in the future is the fuel cell. Like rechargeable batteries, this is an old technology: again, it was invented in the 19th century. These portable electricity-generating devices produce current as a result of a chemical reaction – normally hydrogen combining with oxygen to form water. Unlike ordinary batteries, the chemicals inside them are used up, so

Under the microscope, a forest of carbon nanotubes that make up one electrode of an ultramodern supercapacitor battery. The battery will charge in seconds and last for hours.

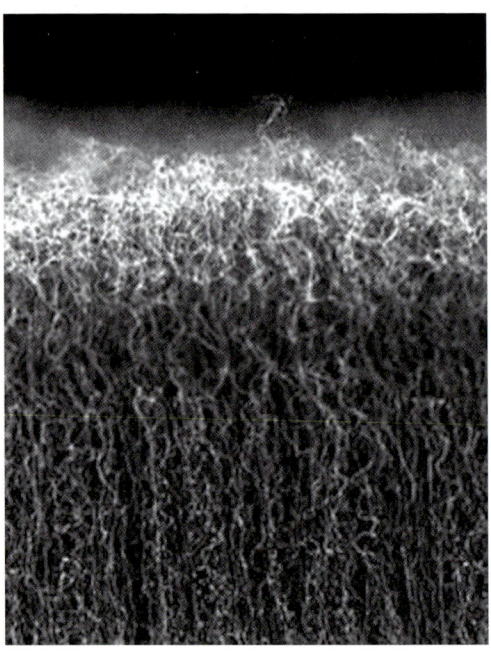

recharging a hydrogen fuel cell means refilling it with hydrogen. The growing likelihood that the world may have to switch to hydrogen as its main source of fuel in the next two decades has prompted a resurgence of interest in fuel cells. And carbon nanotubes star again in some fuel cell designs: the open hexagonal structure of the molecule is ideal for holding onto hydrogen, providing efficient and safe way to store the hydrogen inside the cell.

What we want in consumer electronics we get – eventually. We want more powerful processors – to carry out the necessary computation that will add extra convenience and flexibility to our interactions with our digital information. We want cheap, bright and very thin displays – flexible where relevant. We also want portable devices that we do not have to plug in to the mains every night. And we can have all of this and more. Sometimes we even get things we didn't realise we wanted. Now that the PC is losing its dominance, computing power is increasing and becoming cheaper, displays are becoming better, battery power is less of an issue – and networking is everywhere – are we heading for a techno-utopia? Or is the future slightly less appealing than the naïvely optimistic science fiction stories would have us believe? Or will technological progress halt before we get to that bright future? Are there other barriers besides technological ones that could put a halt to the forward march of the digital revolution?

6 Techno-collapse

Progress in consumer electronics seems to be driven largely by a collective desire to move towards the kind of techno-utopia I described at the beginning of this book. We are living a digital paradigm that is unlikely to shift, and the utopian vision is the natural end point of progress within that paradigm. We seem to be heading along a road to a world in which digital devices do whatever we need them to do. Incredibly powerful portable computers shoved haphazardly in our pockets or embedded cheaply in our clothes? Intelligent machines intuitively serving us information and entertainment at every turn? The decline of the PC, and the keyboard and the mouse, and the rise of effortless ubiquitous computing? Foldable electronic paper that allows us to hold video chats as well as download the latest news? Even though we are not sure of the details, it all feels like a question of 'when', not 'if'. And all the time, progress is accelerating, not slowing down.

The electronics industry is well aware of its mandate to continue carrying us along the road to this kind of vision – and beyond – and is more than happy to profit from fulfilling that mandate. And, on the whole, consumers are willing to pay for being nudged faster and further in that direction. Unexpected developments seem to do little to change our course: incredible but unpredictable innovations are likely to speed our progress along the same road, not send us down a different one. But what if the future doesn't pan out as we expect it to? What if something happens down the road that stops progress or takes us somewhere we don't like? What if, instead of techno-utopia, we encounter techno-dystopia, techno-stagnation or techno-collapse?

Even though we are not sure of the details, **it all feels like a question of 'when', not 'if'**. And all the time, progress is **accelerating**, not slowing down

When Things Go Bad

A major factor in the technological utopia of the future is the fact that our electronic devices will be very reliable, and easy and intuitive to use. You will never have to contact technical help lines or look through instruction manuals; you will never be dissatisfied with the quality of digital media; and all the various technologies will be completely compatible and will work seamlessly together. It is clear that we have a long way to go. Today, we sit at desks watching jerky online video surrounded on-screen by adverts, all delivered through expensive and often unreliable Internet connections. We are frustrated by wireless networks that repeatedly fail us. We have to update our PCs regularly to make sure they are secured against viruses, and be careful what we download with our smartphones. To get the most out of our experiences with gadgets online and off, we need to take time converting files between different formats, learning how to use different software and generally troubleshooting along the way. Programmes on digital television often carry web links or more information directly via an interactive link. The first option is inconvenient – since you have to note down the web address and enter it into a web browser – the second is immediate but always limited in the content it provides. Often, when you download a song or a film, it will not play on one or more of your devices – either because of the digital rights information technology attached to it or because of the variety of different formats, not all of which will play on all devices. The digital landscape at present is far from smooth. And of course there is the expense of keeping up to date with the latest software and hardware. All of these issues can and probably will be solved favourably within ten years. However, eliminating these kinds of annoyances will not automatically bring techno-heaven: there are other, deeper issues to consider.

One of the reasons I have referred to the 'standard model' of the future of consumer electronics as a utopia is that it promises to break down social and economic barriers. It should also improve education and literacy, encourage democracy, and also undermine corruption by leading to transparency in business and government. A fully digitised world filled with cheap, accessible, interconnected and compatible electronic devices does have the potential to achieve these laudable aims. This is very much the premise of several seminal books on the subject of our digital future written in the 1980s and '90s. Of particular note are *The Road Ahead* by Bill Gates and *Being Digital* by Nicholas Negroponte. Some commentators have described the optimistic view presented in these books as 'cyberlibertarianism'. Negroponte talks of four powerful qualities that underlie the 'triumph' of the digital age: decentralisation, globalisation, harmonisation and empowerment. It sounds great, and those qualities truly can be used to the common good. Negroponte was one of the founders of the One Laptop Per Child (OLPC) initiative, that noble venture that attempts to disseminate the benefits that digital technology really can bring – and puts world-changing technology in the hands of the world's most disenfranchised people. But even cyberlibertarians recognise the potential for things to go wrong.

History has witnessed enthusiasm surrounding revolutionary technologies before. Even before the digital revolution began, many writers have declared their belief that 'disruptive' technologies have the power to bring freedom, wealth and a renewed sense of community to everyone. People's faith and hope were placed in the railways, the telegraph and the telephone, radio and television, plastics and nuclear power, to name but a few. But none has delivered utopia – indeed, they have all created problems as well as progress. Often, economics is one of the forces that creates problems surrounding disruptive technologies such as these. Like digital technology, each of those historical technological revolutions has been driven by capitalism, albeit with intervention from governments. The investment needed to produce and implement them is sustained by end-users – consumers. Despite its obvious successes, there are undeniably darker sides to capitalism. This is neatly summed up by a line in the 1962 film *One, Two, Three*: 'Capitalism is like a dead herring in the moonlight – it shines but it stinks.' That stink is largely comprised of greed and the inequity, corruption and massive consumption of materials it can create.

For **now** at least, the Internet remains free of too much **direct control**, and **we are all free to contribute** to it and to choose how and when we use it

Capitalism only works when it produces profits to provide a return on investments. In order to do this, it must result in the consumption of resources and it must maintain a sizable gap between the rich and the poor. Happiness is not necessarily the goal, and is certainly not always the outcome for all; and there have to be winners and losers. In addition, capitalism often concentrates wealth in the hands of a decreasing number of powerful multinational corporations. Increasingly in the digital era, a handful of mega-companies are gaining control of the news, entertainment and publishing industries, which are all converging in cyberspace. Their power over people's opinions and behaviour is enormous – despite the rise of self-made content such as blogs. These companies' interest in our welfare comes after their need for profit, however they might try to convince you otherwise – that is how they survive. They can be more powerful than nation states, and have the potential to control much of the way consumers live out their digital lives.

For now at least, the Internet remains free of too much direct control, and we are all free to contribute to it and to choose how and when we use it. Of course, the relative freedom of the Internet coupled with the increasing power of digital technology poses its own problems. It brings out the worst as well as the best in people. Enter child pornography, identity theft, threats to online security, spam, viruses and serious issues surrounding privacy. As hardware and software become more sophisticated, opportunities for cybercrime will probably grow, not diminish. Unfortunately, whatever efforts are made to counter them will almost certainly affect all of us negatively – in terms of a greater loss of privacy. Attempts to improve our privacy and security as well as our online experience involve the logging of our online behaviours by state organisations and big companies. In a future of ubiquitous computing, the opportunities for unscrupulous governments to carry out surveillance on their citizens will be much greater. China's government already exerts severe restrictions on what sites people there can visit – and on what content its citizens publish. Companies can spy on us too. Small pieces of computer code called 'tracking cookies' already report on our behaviour as we move around the Web, like virtual spies working for advertisers and marketing executives. And our personal details typically exist in hundreds of different locations, where they are out of our control. Add in other factors such as information overload and our bombardment with annoying but supposedly necessary advertising, and the present – let alone the future – seems less than rosy.

Not surprisingly then, just as some writers have expressed their techno-optimism, others have produced convincing scenarios with a less positive spin. Science fiction provides a wealth of visions of technological dystopia. Perhaps the best-known examples are George Orwell's *Nineteen Eighty Four*, which considers the effects of surveillance in a totalitarian state; Aldous Huxley's *Brave New World*, which is concerned with the dehumanisation of people through technological progress; and Ray Bradbury's *Fahrenheit 451*, which is about how the rise of television can lead to the downfall of books. These are all classic novels written between the 1920s and the 1950s, and so they are not concerned directly with the rise of digital technology. Since the 1980s, however, a different crop of dystopian science fiction novels has been produced that does reflect fears about our digital future. The 'cyberpunk' genre features dark portrayals of future high-tech societies gone bad through just the sort of centralised corporate power and corruption I have alluded to. In many cyberpunk stories, much of the action takes place in virtual worlds, and there is almost always a hero who is a computer hacker fighting against the oppressive powers that be. One familiar example is the *Matrix* trilogy, in which robots have dominated human beings and made them prisoners in virtual worlds. Another classic cyberpunk novel is *Snowcrash*, by Neal Stephenson, in which the main character spends much of his time dipping in and out of a realistic virtual world. Cyberpunk stories often include references to ridiculous levels of advertising and surveillance, and a loss of personal freedom that we would find excessive today.

Today's Cyberpunks

Although multinational corporations are not now as powerful as those described in cyberpunk fiction, we already have the equivalent of cyberpunk heroes. There are the communities of programmers that produce free, open source alternatives to restrictive and expensive proprietary software. And there are plenty of activists who campaign against and lobby for the protection of our digital freedoms and protest against the introduction of potentially oppressive technologies, such as radio frequency identity (RFID) tags and digital rights management (DRM).

The future may not be a smooth ride for new technology if that technology interferes with our human rights. There is significant protest against technologies such as RFID and DRM.

We can easily become dominated and controlled by the very technology that we have helped to nurture. The binary digits we send around the world and store and process on our devices are neutral – they are just whizzing electrons, variations in the strength or frequency of radio waves, flashes of light or indentations on the under-surface of optical storage discs. As with anything powerful but inherently neutral, they can be used for good or not-so-good.

Coming to a Standstill

In another, gentler version of our digital future, the consumer electronics revolution simply runs out of steam, and the industry slows down and progress stops well short of its supposed destination. This is a half-baked version of a future not so far down the line. Moore's Law comes to a fairly abrupt end – not through insurmountable technological challenges, but through a lack of available investment for innovation. We simply lose our desire for more, and we stop fuelling the machine. This is a future in which we become satisfied with what we have, and unwilling or unable to shell out yet more money to continue upgrading. There are, after all, so many other ways we have to or want to spend our money.

By **spending our hard-earned cash** on consumer electronics, we may find ourselves **impoverished** in more than monetary terms

At present, we are stuck on a treadmill of progress that could run for decades at least – a state of affairs that many people find frustrating, depressing and ultimately pointless. Indeed, by spending our hard-earned cash on consumer electronics, we may find ourselves impoverished in more than monetary terms – as the Roman philosopher Seneca wrote: 'it is not the man who has little, but the man who craves more, who is poor'. More consumption, more waste, more new technologies to learn how to use, more cost; what is the point of it all? It's like climbing a ladder and realising it has no top – or at least not one that you can see. To continue with this analogy – in the techno-stagnation scenario, we stop to enjoy the view and make sure we have a good foothold, perhaps building a stable platform so that we do not fall.

In this version of the future, the focus is on existing, rather than next-generation, technologies. Smartphones, high-definition televisions and high-definition disc players become commoditised, just as CD and DVD players did before them, and sales eventually slow down to a replacement-level trickle. There will be no improved version to which we can upgrade – we will never be able to buy a PlayStation 25, for example; the devices we buy will not be stepping stones to a more advanced future. In essence, the technology present in this alternative future is an only-slightly supercharged version of today's. The chip factories downsize their operations, because the demand for integrated circuits falls away. Profit margins shrink, and there is little money and no real need to build new multi-billion dollar fabrication plants to make chips with even smaller nano-transistors. The military and scientific communities retain access to their own, smaller-scale, dedicated fabrication plants, so progress does not halt altogether – but the consumer sector slows down to a virtual standstill.

In this future, there is no need to invest hugely in new infrastructure either: in developed countries at least, we will have borne the costs of upgrading to optical fibre networks by 2015. Two gigabits per second turns out to be 'enough'. And we find we don't really need thousands of interconnected MEMS devices scattered around our houses to form *ad hoc* sensor networks. It turns out that we are satisfied with the capabilities of simple, networked portable devices and a multimedia centre in the living room, along with smaller, connected devices in the kitchen and bedrooms. People spend lots of their time online, but the Internet grows only slowly, not rapidly and out of control as it did in its early days. All the groundwork was laid down in developing the Web 1.0 and the Web 2.0 back in the heydays of the 1990s and early 2000s, and the Internet becomes stable and mature – a reliable delivery system for all our communications and entertainment. We still buy content and services – the film and music industries and the network providers, for example, have not been unduly affected. In a sense, this is a kind of techno-utopia – just a diluted version. But the level of investment necessary to sustain progress in hardware, content and services is dramatically reduced. As a result, we are bombarded less by advertising.

The digital revolution is young; it could falter, and it could stagnate – no revolution lasts forever

But is this future at all likely? Hasn't the industry got so much momentum as to make it unstoppable? Well, not necessarily. In ten years, when our homes are adequately networked within and connected to the world at large, and when our portable or wearable devices are as smart as we could ever really need them to be, our interest in new consumer electronics may well start to flag. The industry will have to work hard if it wants to keep us interested indefinitely. The digital revolution is young; it could falter, and it could stagnate – no revolution lasts forever. When LCD and plasma televisions are cheap enough that most of us have them, and we all enjoy the high-definition images they produce, how many of us will be willing or able to upgrade to an OLED screen? And what about the next display technology after that? And how many of us will decide that all we really need from our mobile phone is, well, a phone? Many of us have upgraded our mobiles, say, a dozen times in the first ten years of the 21st century. When that figure reaches twenty or more, might we just think 'that's enough now'? Of course, many people upgrade their mobile phones more because of changing fashions than because of improvements in the technology – but this could still be true in a world in which technological progress stagnates.

There is plenty of evidence that many people already feel this way – and not just the technophobes among us. For example, the 2006 Pew Internet and American Life Project survey, carried out by the Princeton Survey Research Associates, uncovered some relevant findings. Fifty-six percent of respondents stated that they often felt their electronic devices did much more than they needed. Furthermore, only 33% believed strongly that electronic devices made them more productive – and fewer than half really enjoyed the fact that owning a mobile phone made them more available to others.

However, these kinds of attitudes are not prevalent or strong enough to slow things down just yet. Given the thirst we have displayed so far for progress in digital technology, and the promise of better and faster technologies, the revolution is unlikely to come to a halt just now – unless the world experiences a severe and prolonged economic downturn. And even then, it would have to be a severe downturn: in the relatively mild depression of the first few years of the 21st century, the consumer electronics industry managed to survive and remain fairly buoyant. It was helped in part by the allure of thin, flat televisions and personal video recorders, and it was able to produce real innovations and to convince enough people that they needed to buy them – all while keeping prices relatively low and swallowing the burden of minimised profits. The silicon industry's road map was not affected. Clearly, if we have the money to spend, and the gadgets continue to attract our attention, we will keep on buying them for the next several years at least. Expanding further into developing markets such as Africa and South America is likely to be a good economic strategy, too – at least until such a time as consumers there feel they have all they need, too.

A genuine **worldwide** depression or some other thorn in the side of consumer capitalism could **seriously dent people's spending power** and make them more inclined to stick with **what they have**

However, a genuine worldwide depression or some other thorn in the side of consumer capitalism could seriously dent people's spending power and make them more inclined to stick with what they have. There are certainly potential problems ahead for the world's economy. Take oil, for example: most experts believe that we humans are at a moment in our history where the amount of oil left is the same as what we have used, a point that is often referred to as 'peak oil'. In the next twenty years, wars will probably be fought over the dwindling supplies of fossil fuels, and still other skirmishes may erupt over water or other resources. Global warming may well threaten the status quo, potentially displacing millions of people and causing hundreds of billions of pounds-worth of damage. Perhaps one of the first areas of industry that will suffer will be consumer electronics – it may not be business-as-usual for too long.

Hopefully you can at least appreciate the logic and sense in a slowdown. But then, if this slowdown had begun in, say, the 1970s, then we would not have PCs, the

Internet would still be used only by academics, we would have only a few television channels that would not broadcast 24 hours a day, no online banking, only a few brick-sized mobile phones, and we would be wandering around with portable music players with cassette tapes inside them. We would watch films on video cassettes on standard-definition televisions. If we missed a programme on television, we would simply have missed it – unless we remembered to set the video cassette recorder to tape it. We would always be aware that we were living in a world much less advanced than it could be, since back then progress was rapid, as it is today. Then again, that might be the point: progress did not stop in the 1970s because the world was aware of the coming digital revolution, and despite the oil shortages that plagued the early part of the decade, there were no real threats to the economies of the rich countries of the world. In ten years from now, things could be quite different – and the revolution really could be over.

You might feel slightly let down by the idea that the consumer electronics machine that you have helped to build could somehow fail to reach its potential. Where are the carbon nanotube transistors? Where are the teams of household robots? Where are the gadgets that can intelligently serve us? Well, the consolation is that the slow and easy techno-stagnation vision does not do away with digital technology altogether or portray it as an evil; it does not require yet more advertising and loss of freedom. And it is not necessarily a gloomy picture – apart from the fact that it would probably be caused by a depression that would create severe hardship for some people. In this version of the future, there is time to take stock of digital technology, to make our devices safe and secure to use, and enjoy the benefits they can bring – without worrying if the last one we bought is about to become obsolete. In a slowed-down world, technology can get on with the job of making the world fairer and better for everyone, rather than making us playing catch-up and turning us into slaves to progress. In this sober future, digital technology is cheap but not throwaway. There is little disparity between rich and poor countries in terms of the technology they possess and enjoy. In fact, this is perhaps closer to utopia than we are now.

There is a growing movement of **people who reject capitalism**, **globalisation** and the idea of a highly organised, technologically driven **future society**

Back to Basics

To our high-tech dystopian vision and our sober slowdown, we have to add a still more unsettling prospect: techno-collapse. In this scenario, the world is plunged into a more primitive state, with small-scale, locally based physical communities living with just their basic needs. But what kind of forces could bring globalised consumer capitalism to its knees? It has to be more than consumer apathy. Clearly, all-out nuclear war or some other global catastrophe would put a serious dent in our way of life, since it has the potential to destroy our society's infrastructure altogether. There would obviously be no progress in consumer electronics – there would be no chips with 22-nanometre transistors in 2012 if there were a nuclear Third World War in 2010, for example.

Terrorism Against the Future

There is a growing movement of people who reject capitalism, globalisation and the idea of a highly organised, technologically driven future society. Most of these people advocate anarchy in varying degrees, or promote various forms of socialism rather than global capitalism. Since progress in consumer electronics depends upon consumer capitalism, most of those who are anti-globalisation are also against the kind of unbridled progress in consumer electronics and the kind of technocratic civilisation that it may bring. Since the force of technological progress and globalisation is so strong, many of the dissenters feel they are left with no choice but to use force, or even terror, to fight the system.

One such person is Ted Kaczynski, commonly referred to as the Unabomber. In the early 1970s, Kaczynski began living in a remote cabin with no running water or electricity, and lived a hunter-gatherer lifestyle. He became fiercely opposed to the modern technological lifestyle, which he believed would lead to the end of human freedom. He advocated revolution against technology in what he called his manifesto, a 35,000-word essay entitled *Industrial Society and Its Future*. From 1978 until the mid-1990s, he sent bombs to universities and airlines across the USA. He repeatedly demanded that his manifesto be published in national newspapers in return for ceasing his bombing campaign. Kaczynski was eventually arrested in 1996.

However, it is possible that the world will come to be in a post-apocalyptic state without having to suffer an apocalypse. A growing reaction to growing techno-dystopia could bring the downfall of our reliance on modern technology as a way of life; anarchy bringing the system to its knees. The anti-high-tech 'Unabomber' was a single-person operation, but there are others who share the same anti-technology feelings. Already, there are several militant anti-capitalist and antiglobalist organisations. Some of these people are simply fed up with the unfairness of 'the system', and they are hell-bent on stopping the techno-capitalist machine in its tracks and restoring the human race to a more natural existence. What if their number grew exponentially, and they were able to overthrow world order?

The only thing that we can be certain of is change. Nothing lasts forever

This is hard to imagine, and is surely extremely unlikely. It seems that nothing short of apocalypse can bring our way of life to an end. However, perhaps anarchy is unnecessary to put an end to the business-as-usual scenario. There are warnings from history that should be heeded. Even the most powerful civilisations have had their downfall – often relatively quickly after a period of perceived stability. Why should our economic-technological machine be any different? Could there come a time when our high-tech lifestyle requires so much investment on our part that it is no longer sustainable – that we can no longer 'afford' it? Our civilisation could be at a turning point. We are in the midst of an explosive population growth that has led us to a situation in which we require several Earths to sustain our lifestyle. We are retreating further and more often into virtual worlds and we live in a culture of celebrities and sound bites. We are producing so much information that it is easy to lose sight of what we feel is important. All of this could lead to more than simply an unpleasant, unwanted dystopia. It could herald the beginning of our downfall. We humans need various things to survive. Love, community and space to breathe are as important as money, food and shelter. In our modern technological way of life, all of these may soon become scarce. Religious fundamentalism, terrorism, global warming, scarce resources or simply a world that is moving too fast for its own good could just tip the balance and put an end to the reign of our technologically advanced capitalist system.

It might seem that we are wedded to a future in which consumer electronics will play an ever more prominent role – for better or for worse. But not all marriages last – and the only thing that we can be certain of is change. Nothing lasts forever.

7 **Tomorrow's Tomorrow**

In this book, I have laid out the likely progress in consumer electronics in the next few years – perhaps fifteen to twenty. So, assuming a business-as-usual scenario – if consumers and the industry maintain their symbiotic relationship – there will be cheaper, faster and less power-hungry chips and memory; tiny storage devices with near limitless capacities; stunning new displays; new, intuitive ways of interfacing with our digital information; and of course the slow decline of the PC leading us to a new era of ubiquitous computing and an Internet of Things. But what might happen beyond that? Putting aside the question of whether all of this will lead to a technological utopia or not, what kinds of technology might be available further in the future? What are some of the more bold predictions, the outlandish suggestions that lie on the border of science and fantasy? In this last chapter, I will look at what kinds of things might be possible in tomorrow's tomorrow.

Chips With Everything

Today, consumer electronics products are commonplace; most people with even a modest income have several of them. In the far future, electronics will probably

Even the poorest people will have access to **cheap yet powerful** electronic devices, hopefully **empowering them** with access to information and the chance to take part in the increasingly **global economy**

be so cheap and so powerful that, although the gap between rich and poor may well continue to grow, the 'digital divide' may actually shrink. Even the poorest people will have access to cheap yet powerful electronic devices, empowering them, one hopes, with access to information and the chance to take part in the increasingly global economy. Electronic technology has the power to change our everyday lives in a much more fundamental way than just allowing us to send emails, make video calls, and store and access digitised music and video. As prices fall, consumer electronics can and probably will find its way into all kinds of elements of our lives – probably in such a way that the grandchildren of today's children will think us primitive and wonder how we ever coped.

Harvesting Power

Every one of the tiny, low-powered electronic devices that could pepper our environment in the future needs a source of electricity. The amounts of electric power needed will be tiny, and any of several technologies could help to deliver it effectively from the devices' surroundings. Anywhere a difference in temperature exists, for example, there is the potential for generating electric current. This effect was discovered in the 19th century, and is the basis of 'thermocouples' used in electronic thermometers. At a few degrees temperature difference, a thermocouple only generates a few microwatts. However, in tomorrow's world, this may be enough to power a tiny device. For more power, some chips could have energy beamed to them by radio waves, although some governments put limits on the amount of power that can be transmitted in this way. Certain materials, called piezoelectrics, produce electric power when they are vibrated or stretched and squashed. Piezoelectrics in a vehicle could harness enough energy for a low-power device. Even the patter of raindrops on a roof or a window can generate enough power to keep low-power devices going. Researchers at the CEA Leti-Minatec research centre in Grenobles, France, developed a way of using a piezoelectric polymer to harvest energy in just this way.

For our own portable gadgets, embedded in our clothes or handheld, there are other options available – again, if the power requirements are only fairly low. Thermocouples in our clothes could take advantage of the difference in temperature between our skin and the air. There are already devices that can harness energy from the movement of your body. Kinetic watches that mechanically wind a little every time the wearer swings his or her arm have been around for many years. This principle could equally be used to generate electrical power rather than mechanical. Piezoelectric materials in the heel of a shoe could produce electricity, too. And in 2008, scientists at the Georgia Institute of Technology announced that they had developed a fibre made of fibres that generate electricity when they brush up against each other.

All but the most powerful chips – silicon or otherwise – will probably be so small and cheap that many of them will be disposable – and therefore hopefully biodegradable, if developments in organic electronics continue. You will be able to buy thousands of them for the equivalent of a few pounds. Meanwhile, those same chips will probably be extremely efficient – thanks to reversible computing recycling information – and will use virtually no power. And what power they do use, they may even 'harvest' from their surroundings. What kinds of applications might there be for digital processing power and memory that is so small, so cheap, so powerful and yet so frugal in its power needs?

Imagine gaining access to your home or your office **just by touching a door handle**; or sharing information with someone just by **shaking hands**

Well, firstly, our own 'personal area networks' may become a way of interacting with chip-enabled objects and people in the world – through touch. Imagine gaining access to your home or your office just by touching a door handle; or sharing information with someone just by shaking hands. Imagine dictating a document while you are walking home and then simply touching your electronic book when you get home and watching your words appear on the page. Touch technology could help elderly patients to ensure they take the correct medicine: when they open a bottle of pills, a chip embedded in the bottle would recognise the patient and note how many of the pills he or she had already taken. In a shop, imagine picking up a product and then having information about it displayed immediately on your handheld information device, or perhaps a virtual retinal display that projects information directly in to your eye.

All of this could be possible by having a suitably equipped device somewhere on your person – perhaps even a chip embedded under your skin. The device would communicate by sending a weak electromagnetic signal along the surface of your skin. Strange though this idea may sound, this touch-mediated 'human area networking' technology has been the subject of research efforts for more than a decade. In 2007, a Japanese company called the Nippon Telegraph and Telephone Corporation demonstrated a working version, called RedTacton – and it works rather well. The transmitting device initiates a fluctuating electric field that carries coded digital information along the surface of the skin. The receiving device contains a crystal whose optical properties change with a fluctuating electric field, and uses a laser to probe those properties and retrieve the information. The electric current passing along the skin is minuscule, and poses no threat to health. Real-world applications should be available in a few years' time. Researchers working on RedTacton have already achieved simultaneous transmission and reception speeds of up to 10 megabits per second. The difference between this and wireless communications technology is that a connection is established whenever your skin comes into contact with a device, and is terminated automatically when that contact is no longer present. This could be an advantage in certain applications.

Another application of practically free chips not too far in the future will be to extend the importance of RFID (radio frequency identification) in creating a ubiquitous Internet of Things. Simple identification chips could be attached to any and every product – from food, luggage and clothing to cars and lorries – and of course, our electronic gadgets. Each chip would be unique and would have some built-in intelligence and the capability to form *ad hoc* networks – they would be addressable in the Internet of Things. By combining these identification chips with satellite and other location systems, we would always be able to find our objects

In the future, a touch could say so much.

if they were lost or stolen: they could transmit their exact position, and we could search them online. Tracking services like this are already available for cars and lorries, via GPS and radio transmitters. But in a future age of ubiquitous computing, it could be used practically free of charge with any number of our belongings. This would generate unthinkably large amounts of information, and it raises a number of important questions about privacy – but it could prove useful if and when it becomes possible.

Simple identification chips could be attached to any and every product – from **food**, **luggage** and **clothing** to cars and lorries – and of course, our **electronic gadgets**

In the future, your lost shoes could let you know their whereabouts.

The idea that even our most trivial belongings could be 'geolocative' – findable anywhere on Earth – is most closely associated with science fiction writer and futurologist Bruce Sterling. According to Sterling, instead of looking around for our shoes in the morning, we may simply 'google them'. However, Sterling points out that the importance of these objects goes much further than simply having things you can't lose – although that would be useful. He suggests that they could help our consumer society be a little bit more green. Every piece of information about every object – when and where it was made, and from what, who owns it and where it is now – would be available online. This would allow consumers – or their intelligent software agents – to track an object through its lifetime, and would make it easy to dispose of it or recycle it in the most responsible and best possible way. In his book *Shaping Things*, Sterling names this kind of object as a 'spime' – a contraction of 'space and time', since these objects are locatable in space and trackable in time. Sterling suggests that the most important role for these objects will be in helping us to achieve sustainability, by 'cradle to cradle' production. When objects are trackable throughout their lifetimes, and have information about what they are made of and where they come from immediately available, there should be much less tendency to dispose of objects in the traditional way. There may even be whole industries that grow up to salvage spimes. Artist and technologist Julian Bleecker uses the term 'blogject' to encourage the idea that objects would have the ability to 'communicate' by posting information about themselves or their environment as a kind of blog.

Another way in which extremely low-cost but very powerful chips may manifest themselves is in distributed intelligence. Home appliances, vehicles, and toys will probably have chips for sensing and processing that give them sufficient levels of intelligence to act autonomously and intelligently. Perhaps we will be routinely holding conversations with inanimate objects. In fact, the chips may be inside our brains – for one area in which there is likely to be significant progress is in neural implants. The impetus for this kind of research is the desire for treatments for stroke victims whose brains have suffered severe damage in particular areas or for patients suffering paralysis. Already, neural interfaces that detect specific activity in patients' brains have been used successfully to control external robot arms and to move a computer cursor by the power of thought alone. In 2008, the activity of just two hundred neurones in the brain of a monkey on a treadmill at Duke University, in North Carolina, USA, controlled a pair of robot legs walking on a treadmill at the Computational Brain Project in Kyoto, Japan. The monkey had visual feedback of the Japanese robot, and was able to keep the robot walking just by thinking about it – even after its own treadmill was switched off.

Researchers have been able to make good electrical connections between neurones and silicon-based circuits since the 1990s. In 2006, a group led by Charles Lieber at the Massachusetts Institute of Technology grew rat neurones on top of a silicon chip crossed by 20-nanometre-wide wires. As a result, the researchers effectively hooked up the inputs and outputs of the neurones – the axons and dendrites – to the individual nanowires. They were able to monitor and interact with the neurones' electrical activity.

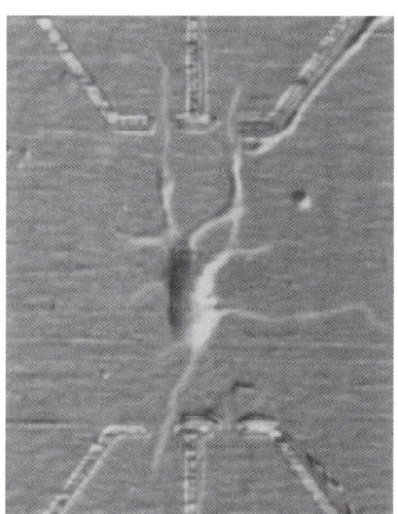 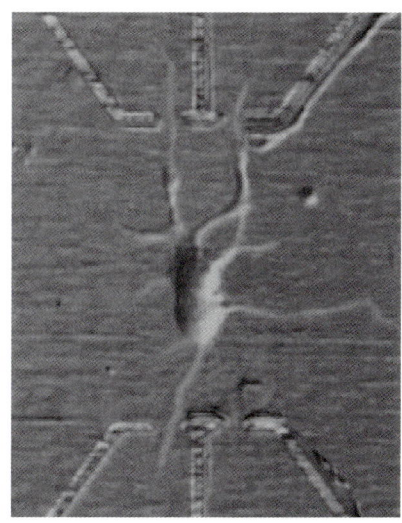

Further into the future, some of us may actually **elect** to have **neural** implants – because **they may enhance our own human powers** of memory or perception, not simply make up for **deficits** caused by **accident or disease**

Electronic devices that can interface directly with the brain like this have the potential to make paralysed patients mobile again by controlling their muscles. Amputees could control intelligent prostheses. In that sense, such devices will be consumer electronics products – but fortunately with a limited market. However, further into the future, some of us may actually elect to have neural implants – because they may enhance our own human powers of memory or perception, not simply make up for deficits caused by accident or disease. Would you choose to have a 'bionic' eye that could display information directly in your field of vision, might have a zoom function and would connect wirelessly to the Internet of Things – if it was cheap, could be installed quickly with a painless operation? How would you feel about having an eye that was both a consumer electronics item and a spime? The area of 'bio-informatics' and 'wetware' is likely to be a major driver of technological change in the next few decades.

It's Not About the Hardware

All this talk of 'chips' may seem very outdated in twenty years: even the term 'electronics' may be obsolete, since it refers to devices that control the flow of electrons. Future devices may not use electrons at all. Dutch computer scientist Edsger Dijkstra once said that computer science is as much about computers as astronomy is about telescopes. Abstracted from its hardware, computing is about information. Anything that can represent and manipulate information is a possible technology for the future of computing, and therefore potentially in our consumer gadgets. The consumer electronics revolution – and computing in general – is not about the hardware. What defines an area of technology and drives it forward is not simply pushing the boundaries of what is possible – although that clearly plays a part. Instead, it is people's needs and desires: what people want, or least what they think they want. So, in the end, the future of information technologies is about what those technologies can do for us and what we want them to do. Communication, information and entertainment.

Every great 'disruptive' technology has its day, and electronics as it stands today will give way eventually to something else that does the same thing better. Even the internal combustion engine may well be superseded in the next twenty years, as a result of environmental pressures and the development of better alternatives. However, people will still want to travel. In the same way, PCs may give in to ubiquitous computing, and that may develop into a quite different phenomenon afterwards – but people will still want to create, store, process and access information.

Binary digits can be **represented** by any two-state **system** – not only by **electric currents or pulses** of laser light

Binary digits can be represented by any two-state system – not only by electric currents or pulses of laser light. Anything that can handle binary information more quickly, more cheaply and more effectively than silicon-based or carbon nanotube-based electronics will take over eventually. Information need not even be binary: it could be based on a system with 8 or 16 bits, or even more. In fact, information processing in the far future may not even be digital at all. It is possible to process information using 'analogue' signals – quantities that can take any value, not just '0' or '1'. Continuously varying electric current or light intensity may not sound ideal for representing information, especially after reading here about the versatility of digital electronics. But analogue computing may prove to be faster and more suitable for creating artificial intelligence than digital computing can ever be. The brain is more like an analogue system than a digital one. Although a neurone (brain cell) is normally described as on (firing electrical pulses) or off, there is actually a range of firing rates and the response to and interaction with other neurones is far from two-state. Digital computing is actually rather clumsy when compared to the analogue approach; digital circuits are more complex, more power-hungry and larger than an analogue computer can be. For now, the versatility of the digital approach will win through. But some researchers, such as Jonathan Mills at the University of the West of England, believe that analogue technology will eventually supersede digital technology. Mills has designed and built several analogue computing chips that work very well for controlling movement in robots and as sensors in robot vision.

We can only guess what the hardware of the far future may be like and how it might work. There are a few favoured technologies, which exist in the minds of theoretical physicists or as limited experimental set-ups, but which may one day come into the mainstream. They will only do so if they can outperform the silicon- or carbon nanotube-based processors and memory chips that we have or will have. And if these technologies take off and become practical, they do indeed promise to take the Information Age to another level altogether.

There are a few favoured technologies, which exist **in the minds** of **theoretical** physicists or as **limited** experimental set-ups, but which **may one day come into the mainstream**

The Future of Bits

All of our modern consumer electronics gadgets are digital devices that use binary numbers to represent information. The emerging field of spintronics, using electrons' spins to represent information, rather than their electric charge, may change that. By freezing an electron's direction of spin as it rotates passing through a spintronic device, it is possible to leave the spin in any of several distinct orientations. One idea is to use this variety of spin orientations as the basis for digitising information. Since each separate orientation is frozen as a different phase of the passage of an electron through a device, researchers studying the potential of this technique call them 'phits' (phase bits) rather than bits (binary digits). If this idea ever comes to fruition, devices would use sixteen phits instead of the customary two bits (0 and 1). You could represent any of the numbers from 0 to 255 using only two phits, compared with eight bits in the binary system. The advantage is that information could be processed and communicated at much higher rates.

Quantum Information

The first of the potential successors to integrated circuits as we know them today is quantum computing, which makes use of the weird behaviour of matter and energy at the atomic and subatomic level. One of the strange properties of quantum systems such as an atom is that they can exist in up to an infinite number of different states at once – a phenomenon called quantum superposition. While

In the future, digital devices may use a number system based on sixteen rather than on two. To see how this might be possible, consider a spintronic device in which the electrons' spins, first aligned in one direction, can change incrementally to any of sixteen orientations. Ditching the binary system would speed things up considerably.

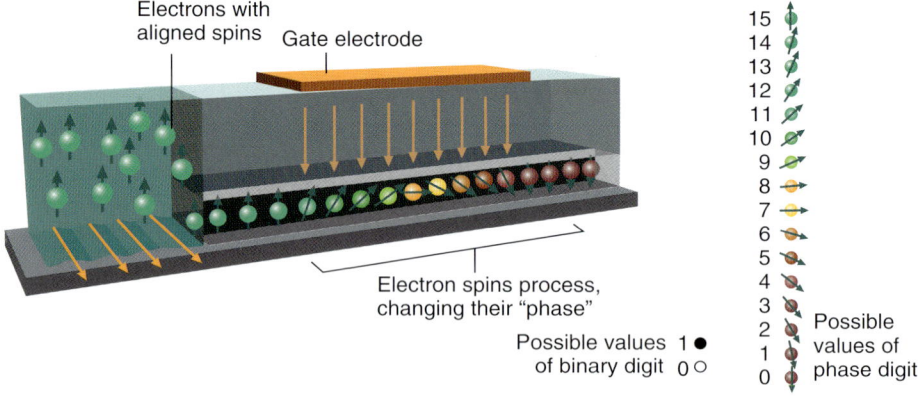

Electrons with aligned spins   Gate electrode

Electron spins process, changing their "phase"

Possible values 1 ●
of binary digit 0 ○

15
14
13
12
11
10
9
8
7
6
5
4
3
2   Possible
1   values of
0   phase digit

conventional binary computing uses binary digits, which may take either of two values, quantum computing uses quantum bits, or qubits to represent information. Thanks to quantum superposition, a qubit can have an infinite number of values simultaneously. Qubits can be represented by photons ('particles' of light), electrons or whole, individual atoms.

Quantum superposition is the basis of the well-known Schrödinger's Cat Paradox – a thought experiment in which a cat is held in a box with a sealed vial of poisonous gas. The vial will be broken if a radioactive atom decays; and that decay process obeys the laws of quantum physics. According to those well-tested laws, the atom exists in both its undecayed and decayed state until someone opens the box and looks inside. It is not just that an experimenter would not know what happens to the atom – and therefore the cat: until the act of observation takes place, the radioactive atom really exists in both states. This is weird and counter-intuitive, but it really is so. Austrian-Irish physicist Erwin Schrödinger designed this paradox in 1935 to highlight the problem of interpreting the strange world of quantum physics. The superposition of states of the radioactive atom is not in question – that can be demonstrated in a number of different ways. It is just that it leads to an absurd fact: the simultaneous life and death of the cat.

It is the **'most likely' answer** that emerges at the end of a quantum **computation**. This would be a **major shift** in what 'computing' means

Qubits, in a suitably set-up experiment, can represent all possible answers to a mathematical problem at once. The answer emerges when the system is observed. One of the foibles of quantum computing is that it relies on probability, just as every quantum system does – there is a fifty-fifty likelihood that the radioactive atom will decay in Schrödinger's Cat Paradox, for example. And so, it is the 'most likely' answer that emerges at the end of a quantum computation. This would be a major shift in what 'computing' means.

A quantum computer with just a few tens of qubits would have millions of times the processing power of a conventional supercomputer. So far, researchers have designed and run quantum computers that use up to 16 qubits. But these devices do not look like any computer you would recognise: they are run in physics laboratories on laser benches with mirrors and lenses; or with 'ion traps', which hold individual, electrically charged atoms in magnetic bubbles in vacuum chambers. Alternatively, they run in sealed apparatus cooled to within a tiny fraction of a degree of absolute zero or in chambers surrounded by powerful magnets that produce nuclear magnetic resonance – the phenomenon used in MRI brain scanners. These sophisticated but at the same time primitive quantum computers have been used to solve extremely difficult mathematical problems in a very short time. For example, quantum computing has cracked tough encryption codes – the sort of problem that can be done with conventional computers, but only over a long period of time and with large amounts of computing power.

However, the practical problems of quantum computing are great, and it could be decades before a quantum computer comes into the hands of consumers. It may be that quantum computers only find applications in scientific research or in cryptography (code making and code breaking). There have been many proof-of-concept demonstrations, but any real-world applications are very far off – despite the fact that hundreds or even thousands of researchers are working hard to push the science and technology along. Ultimately though, there is no reason why quantum computing cannot be at the heart of the next information revolution. We are, perhaps, at a similar stage of development as the early computer pioneers were in the 1930s and '40s. They knew from their theories that digital electronic computers should work, but the techniques and technologies for making them took some time to figure out – and powerful handheld computers did not become a reality until a long time after that.

One of the problems with quantum systems is that they are very sensitive to their surroundings – that is why they have had to be built in vacuum chambers or at such cold temperatures. For example, a qubit based on the spin of an electron – spin is a quantum property – can be affected by the spins of other electrons and even the spins of nearby atomic nuclei. In 2008, a team at the Solid State Physics Laboratory in Zurich produced a type of structure called a quantum dot, which can contain just a few electrons and isolate them from such influences. They made their quantum dot from a tiny piece of the material graphene – the first time a

The supercomputer of tomorrow's world?

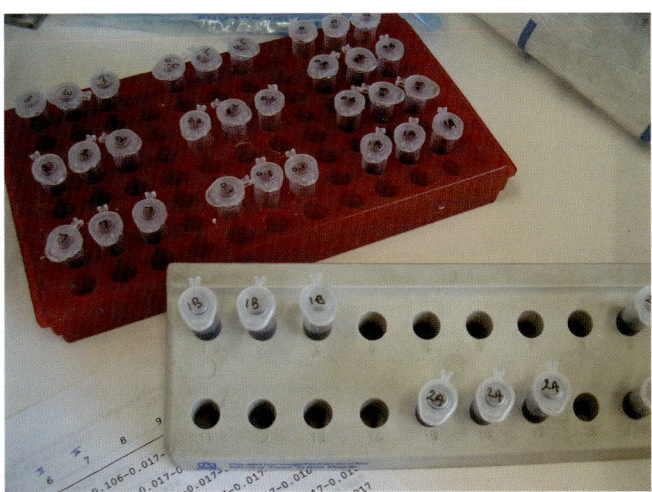

quantum dot has been made from this material – and they hope to work out ways to use this kind of arrangement to trap single electrons, manipulate their spins and use them to carry out quantum computation. This is the kind of development that could help to enable quantum computing to come out of the laboratory and into physical devices. But still there is a long way to go.

Molecular Messages

Another radically different way of representing and manipulating information is chemical computing. This is at a more advanced level of development than quantum computing, but is still a very long way off any really useful applications

In a chemical **computer**, information is represented by **concentrations of chemicals** and **reaction** rates

– particularly anything that could find its way into consumer products. And at present, as for quantum computing, it is a very different concept from conventional computing. There are no wires and switches: just chemicals reacting together. Chemical reactions can be seen as computations – they are nature's way of calculating what would happen if you add chemical A to chemical B, for example. So, in a chemical computer, information is represented by concentrations of chemicals and reaction rates. In recent years, chemists have been able to design chemical logic gates – the basic building blocks of computer processors – from specially chosen molecules. Again, if researchers can harness the information processing capability of chemical reactions in a practical way, chemical computing has great potential.

A variation of chemical computing is biomolecular or DNA computing. The DNA molecule is composed of two complementary strips of chemical units called bases. There are four different bases, referred to as A, G, C and T (for adenine, guanine, cytosine and thymine), and they are complementary in that they 'fit together' in only two combinations: C-G and A-T. As such, they hold the two strands together something like a zip. The four bases form a code that holds information and executable 'instructions' for making proteins, the building blocks of living things. And so, as you can see, DNA is really just a means of carrying information, and biochemical chemical reactions between enzymes and strands of DNA are a form of information processing. In DNA computing – for now at least – there is no keyboard and mouse, no display screen or printer. Instead, DNA computers are test tubes filled with clear solutions. The DNA strands are in the solution and carry the 'input', encoded as engineered sequences of bases. The output is in the form of modified DNA strands that must be analysed using techniques commonly found in genetics laboratories. Because of the size of DNA molecules, a tiny drop of solution can contain trillions of computers all working at the same time.

DNA computers may find their greatest application as **programmed medical 'nanorobots'** that can engineer **conditions inside our cells**

As with quantum computing, DNA has so far been used only to solve specific and well-defined mathematical problems – albeit very complex and difficult ones. In 2002, researchers from the Weizmann Institute of Science, in Israel, designed a programmable molecular computing machine composed of proteins and DNA molecules that could perform 330 trillion operations per second – much faster than the world's fastest supercomputer. This 'device' was the first programmable computer in which the input, output, software and hardware were all made of DNA.

In the far future, there may be devices that would use ordinary integrated circuits for normal information processing, but that would have DNA 'processors' that can take over specific tasks for which they would be better suited. So, although there is hope that DNA and molecular computing might one day be incorporated into consumer devices for storing and processing information, it is more likely that they will have uses that do not fit into the mould of 'consumer electronics' as we understand it today. DNA computers may find their greatest application as programmed medical 'nanorobots' that can engineer conditions inside our cells in response to the first stages of cancer or other intracellular diseases.

I Am Machine

One of the main advantages of quantum, chemical and DNA computing is that they have the power to carry out huge numbers of calculations simultaneously. It is no coincidence that this massive 'parallelism' is also a feature of our brains. After all, we are made up of atoms and molecules and DNA, although that is not to say that the human brain actually uses quantum computing or molecular computing to carry out its 'calculations' – no one knows yet. However, with such enormous processing power at our disposal in future technologies – even silicon-based ones – might we be able to match the power of the human brain? There is little sense in making direct comparisons of brains and raw computing power: our understanding of how brains work has improved hugely in the past decade, but we are still a long way off having a working theory of how the brain works. Brains are certainly very different from the operation of conventional computers – apart from artificial neural networks that aim to mimic the way brain 'circuits' work. Creating true artificial intelligence is more difficult than simply turning up the processing power.

If **Moore's Law** continues on course, we will hit that mark within about 20 years: a portable device **with the power of our brains** by **2030**?

However, when and if we understand the brain well enough, and are able to build information processing devices that can work along similar lines, we will no doubt need enormous processing power to make them produce results similar to those produced by our brains. The lump of interconnected neurones inside your skull receives billions of inputs from your senses, and somehow produces sensible output in terms of thoughts, intentions and the coordinated movement of muscles. There is a great deal of evidence that what the brain is doing is carrying out computation: for example, even digital computers can model accurately the behaviour of individual (analogue) neurones. Despite the theoretical and practical challenges in working out the computing power of the human brain, several information theorists have attempted to make rough estimates, in terms of calculations per second, by analysing how much information specific groups of neurones receive and output or by accurately modelling the action of specific regions of the brain using artificial neural networks. The estimates range from around 100 million million calculations per second up to around a million million million. This is only between 1,000 and 100,000 times more powerful than the most powerful supercomputers in existence today. If Moore's Law continues on course, we will hit that mark within about 20 years: a portable device with the power of our brains by 2030? What kinds of applications could that have?

Artificial intelligence is set to revolutionise our everyday lives long before computers can be built with that kind of processing power. With a computer even as powerful as a dog's brain running the appropriate software, devices have the potential to become as intelligent and as useful as guide dogs and sniffer dogs, but with access to all the world's available information. With advances in the hardware of robotics – such as artificial muscles and artificial senses – this intelligence could be the brain of a trusted and faithful servant that could also act as a conduit for information and communication. Perhaps these machines could be nurses, cleaners or home helps – all of these applications are already being tested, even with the limited intelligence of today's robots. However quickly and however far artificial intelligence moves ahead, robots will soon be so cheap and versatile and intelligent that they could be as ubiquitous as computers.

When and if artificial intelligence evolves to a level that compares with our own, questions of ethics will become very important. Will computers be conscious? Will they have free will? Will devices made from non-living materials be thought of as living beings? And if so, will we have to see them as partners in our future rather than servants? And of course, if devices with intelligence equivalent to the brain become reality, technological progress will not stop there. What lies beyond? Technological progress has been exponential: most indicators of that progress double every two years, meaning that that rate of progress increases. So perhaps by 2040, portable devices will have intelligence equivalent to the entire human race. And in 2050? While such enormous intelligence could help to overcome many of the major problems facing the human race, the possibility does make you wonder about the future of humanity. Would our 'devices', by then far superior to ourselves, need us any more?

The point at which artificial intelligence reaches the same level as the human brain – hypothetical though it may be – is referred to as the 'Singularity'. The term was coined by science fiction writer Verner Vinge in the 1980s, although the concept had been discussed as early as 1965, by statistician Irving Good. Vinge's intention in devising this term was to compare what may happen to human society when artificial intelligence progresses beyond human intelligence with what happens near the centre of a black hole, a point called a singularity. Around a physical singularity, the normal laws that govern space and time break down; perhaps something similar may happen to human civilisation when machines become more intelligent than their creators. However, many commentators suggest that the Singularity will not mean the end of human civilisation nor even necessarily be a bad thing. Computer scientist Ray Kurzweil is one of them. In his book *The Singularity Is Near*, Kurzweil looks in detail at the trends surrounding Moore's Law, and considers the kinds of extreme technologies I have discussed here. He suggests that the Singularity is an inevitable next stage in human evolution, and that we shouldn't fear it.

Along the way to the Singularity – assuming we can develop a workable theory of the human brain – artificial intelligence combined with neural implants could lead to a form of telepathy. Direct brain-to-brain communication mediated by wireless networking, or perhaps by something like RedTacton touch technology; virtual reality gone crazy, with completely immersive multi-sensory virtual worlds or video games pumped directly into our brains – just like being there. Perhaps we will be able to 'download' and share our memories. This kind of 'cybernetic' vision is found in the pages of science fiction novels, and it is certainly a far cry from the media players, smartphones and PCs that are characteristic of today's consumer electronics. But science and technology are advancing at such a rate that they might be possible one day – perhaps even in your lifetime.

It's Not All Good

So, even your socks would have a web page? That's information overload gone mad. Of course, we would not have to deal with that kind of information directly. But crazy information overload is not the only potential problem in an information-rich future. I have already alluded to the possibility of our techno-utopia going rotten and becoming a techno-dystopia instead, so I will not dwell too long on the scary possibilities. But with computing power so readily available and all around us, and given the fact that any technology can be used both for good and for bad, the possible downsides of advanced technology in the far future deserve serious

The **threats** could be **much more** serious than we imagine, and the **ultimate losers of the game** could be **consumers** themselves consideration. With so much information and communication flying around, and so much reliance on information technology, the threat of spam, viruses – and other kinds of dangers as yet unknown – is likely to increase. Until now, antivirus and anti-spyware experts are able to update their protective software as soon as a new threat arises, and savvy consumers keep up to date, minimising the potential damage. However, nobody knows whether in the future, the disruptors or the protectors will get ahead of the game; the threats could be much more serious than we imagine, and the ultimate losers of the game could be consumers themselves. And the more we come to rely upon information technology in our everyday affairs, the more we have to lose if it all collapses as a result of cyberterrorism or catastrophe.

There are other phenomena of a darker nature that may feature in a far future of cheap, ubiquitous, interconnected information processing products that have embedded intelligence. For example, these devices will increase the likelihood and the effectiveness of surveillance. Spy technology in the workplace could monitor our every move and our productivity. Police would be able to keep a close eye on suspects without them knowing. Add to this kind of fear the chance that post-Singularity, hyper-intelligent robots may make our species redundant, and there really is a lot to think about. Is this what I unwittingly support whenever I buy a new television or upgrade my mobile phone? Should I live in a cabin and become a hunter-gatherer – though without the terrorism?

Consumer **electronics** or consumer information technology – **whatever form it might take** – just might help to make tomorrow's world into a **techno-utopia** after all

The potential for progress in the technology underpinning our information age is certainly dizzying. Much of it may well be fantasy – many of the predictions of even informed science fiction writers from the 1950s and '60s have turned out to be way off. Nevertheless, progress is certainly rapid, and who knows where it might lead? Some of the potential applications are negative, scary even – so should we stop buying consumer electronics products? If we do, progress will slow down – which might not be a bad thing – and the impact on the environment will be greatly reduced. However, the digital revolution is still young and incoherent, and it does hold the potential for doing good. As with any disruptive technology, it is up to those who make it, those who use it and those who govern to steer it in the right direction. For consumer electronics to progress, consumers need to keep buying products to fuel research and development. Perhaps consumer power and political power can encourage manufacturers to use an ever greener approach, including making low-power, recyclable and reusable devices. Of course, this is most likely to emerge as the result of economic incentive – from environmental pressures and perhaps the new opportunities arising from the salvage of spimes. Despite the undeniable power of capitalism, however, our species somehow, collectively, needs to make sure that it steers the world to a fair future. And somehow, it must continue to narrow the digital divide, ensuring that everyone who wants it has access to the latest technologies that will allow them to communicate, find and exchange information, and to entertain and be entertained. That way, consumer electronics or consumer information technology – whatever form it might take – just might help to make tomorrow's world into a techno-utopia after all.

Wherever technological progress leads us in twenty, thirty or even fifty years, it is important to remember that the future is not a destination – it does not end there. Tomorrow never comes.

Index